# Mining Multimedia Documents

# Mining Multimedia Documents

Wahiba Ben Abdessalem Karaa and Nilanjan Dey

CRC Press
Taylor & Francis Group
Boca Raton London New York

CRC Press is an imprint of the
Taylor & Francis Group, an **informa** business

CRC Press
Taylor & Francis Group
6000 Broken Sound Parkway NW, Suite 300
Boca Raton, FL 33487-2742



Printed on acid-free paper

International Standard Book Number-13: 978-1-138-03172-2 (Hardback)



**Library of Congress Cataloging-in-Publication Data**

Names: Karaa, Wahiba Ben Abdessalem, 1966- editor. | Dey, Nilanjan, 1984- editor.
Title: Mining multimedia documents / edited by Wahiba Ben Abdessalem Karãaa and Nilanjan Dey.
Description: Boca Raton : CRC Press, [2017] | Includes bibliographical references and index.
Identifiers: LCCN 2016051050| ISBN 9781138031722 (hardback : acid-free paper) | ISBN 9781315399744 (ebook) | ISBN 9781315399737 (ebook) | ISBN 9781315399720 (ebook) | ISBN 9781315399713 (ebook)
Subjects: LCSH: Multimedia data mining. | Content-based image retrieval.
Classification: LCC QA76.9.D343 M54 2017 | DDC 025.040285/66--dc23
LC record available at https://lccn.loc.gov/2016051050

**Visit the Taylor & Francis Web site at**
**http://www.taylorandfrancis.com**

**and the CRC Press Web site at**
**http://www.crcpress.com**

MIX
Paper from responsible sources
FSC www.fsc.org FSC® C013056

Printed and bound in Great Britain by
TJ International Ltd, Padstow, Cornwall

# Contents

# Preface

## Objective of the Book

Nowadays, a huge amount of data is available due to the advances in information technology (IT). In this Information Age, information has become much needed and easier to access. High digitalization of information, declining costs of digital communication, increased miniaturization of mobile computing, etc., contribute to the high demand for information. Also, the progress made in the multimedia domain allows users complete access to digital information formats (text, image, video, audio, etc.).

Most users and organizations need to handle multimedia documents. For this purpose, a large number of techniques have been proposed, ranging from document processing—acquisition, collection, storage, formatting, transformation, annotation, visualization, structuring, and classification—to more sophisticated multimedia mining documents, such as automatic extraction of semantically meaningful information (knowledge) from multimedia documents.

The development of the Internet, also, has made multimedia repositories huge and widespread. There are many tools and methods to search within this large collection of documents, but the extraction of useful and hidden knowledge is becoming a pressing need for many applications and users, especially in decision making. For example, it is of utmost importance to discover relationships between objects in a medical document based on the variety of content. The document can be a medical report that contains a description of medications administered to a patient and scanned or MRI images showing the progress of the patient. Images can be mined, integrating information about patient treatment and patient condition. Extremely important relationships between drugs and disease can be revealed based on image-processing techniques and, at the same time, on natural language processing (NLP) techniques.

*Mining Multimedia Documents*, as the title of this book insinuates, is a combination of two research fields: data mining and multimedia. Merging the two areas will promote and advance the development of knowledge discovery in multimedia documents. It responds to the increasing interest in new techniques and tools in multimedia disciplinary, such as image analysis and image processing, and also techniques for improving indexation, annotation, etc. At the same time, it responds to the increasing interest in advanced techniques and tools in data mining for knowledge discovery. Multimedia document mining is an area that still has scope for development.

## Target Audience

This book represents an investigation of various techniques and approaches related to mining multimedia documents, considered today as one of the most outstanding and promising research areas. This book is a significant contribution to the field of multimedia document mining as it presents well-known technologies and approaches based on text, image, and video features. It also provides an important insight into the open research problems in this field.

The book will also be helpful to advanced undergraduate students, teachers, researchers, and practitioners who are interested to work in fields such as medicine, biology, production, education, government, national security, and economy, where there is a need to mine collected multimedia documents.

## Organization of the Book

The goal of this book is to reassemble researchers in data mining and multimedia fields. It presents innovative researches along the three sections dealing with text mining and multimodal document mining. The book is organized into 15 chapters. A brief description of each of the chapters follows.

Chapter 1, "Mining Multimedia Documents: An Overview," focuses on real-world problems that can involve multimedia mining and proposes a literature review of approaches dealing with multimedia documents, taking into account various features extracted from the multimedia content. It distinguishes between static and dynamic media. The multimodal nature of multimedia data creates a need for information fusion for segmentation analysis, indexing, and even retrieval.

Chapter 2, "Fuzzy Logic for Text Document Clustering," denotes that fuzzy logic has become an important field of study thanks to its ability to help researchers to manipulate data that was not accurate and not precise. This chapter proposes an approach based on fuzzy logic and Euclidean distance metric for text document clustering. The idea is to search for the similarities and dissimilarities between biological documents to facilitate the classification task.

Chapter 3, "Toward Modeling Semiautomatic Data Warehouses: Guided by Social Interactions," is aimed at modeling data warehouses that are used to support decision-making activities in systems of business intelligence to ensure the structuring and analysis of multidimensional data. The chapter proposes a novel approach to design data warehouses from data marts based on a descriptive statistics technique for the analysis of multidimensional data in the principal components analysis (PCA) framework in medical social networks.

Chapter 4, "Multi-Agent System for Text Mining," gives an overview of text mining concepts and techniques applied to extract significant information from a text. The chapter focuses on the application of the paradigm multi-agent systems (MAS) applied generally to distribute the complexity among several autonomous entities called agents. The main objective of this research is to indicate the applicability of MAS technology to find adequate information from texts.

Chapter 5, "Transformation of User Requirements in UML Diagrams: An Overview," focuses on the process of extraction of Unified Modeling Language (UML) diagrams from requirements written in natural language. This chapter provides a survey on the transformation of requirements into UML diagrams and a comparison between existing approaches.

Chapter 6, "Overview of Information Extraction Using Textual Case-Based Reasoning," attempts to support the idea of information extraction that can be performed to extract relevant information from texts using case-based reasoning. The chapter provides an overview of some approaches to illustrate this idea. It also presents a simple comparison of some systems that use textual case-based reasoning for information extraction.

Chapter 7, "Opinion Classification from Blogs," discusses blogs that accumulate large quantities of data that reflect user opinion. Such huge information is automatically analyzed to discover user opinion. In this chapter, a new hybrid classification approach for opinion (CAO) from blogs is presented using a four-step process. First, the dataset from blogs is extracted. Then, the corpus is processed using lexicon-based tools to determine the opinion holders. Thereafter, the corpus is classified using a new proposed algorithm: the Semantic Association Classification (SAC). The generated classes are finally represented using the chart visualization tool. Experiments carried out on real blogs confirm the soundness of the proposed approach.

Chapter 8, "Document Classification Based on Text and Image Features," presents an approach for multimedia document classification. This approach takes into account the textual content and image content of these documents. The idea is to represent a document by a set of features to improve classification results. This chapter explores the state of the art in document classification based on the combination of text features and image features. It also evaluates various classification methods and their applications that depend on text-image analysis, discusses the challenges in the field of multimodal classification, and proposes some techniques to overcome these challenges.

Chapter 9, "Content-Based Image Retrieval Techniques," discusses the most extensively used image-processing operation. Content-Based Image Retrieval (CBIR) aims to reduce complexity and obtain images correctly. The authors show that image retrieval depends on the fitting characteristic extraction to describe the coveted contents of the images. They indicate that CBIR is a context that retrieves, locates, and displays most visually similar images to a specified query image from an image database by a features set and image descriptors.

Chapter 10, "Knowledge Mining from Medical Images," deals with the extraction of convenient information from image data in medicine and the health sciences. A research work as a cutting-edge in relevant areas was presented. This was done to fill the gap for evolving medical image databases instead of simply reviewing the present literature. This chapter initiates a discussion for the data mining and knowledge discovery and data mining (KDD) context and their connection with other related domains. A recent detailed KDD real-world applications summary is offered. The chapter includes a variety of methodologies and related work in the medical domain applications for knowledge discovery. Furthermore, it addresses numerous threads within their broad issues, including KDD system requirements and data mining challenges.

Chapter 11, "Segmentation for Medical Image Mining," introduces the image mining concept in the medical domain. It represents a survey on several image segmentation methods that were suggested in earlier studies. Medical image mining for computer-aided diagnoses is discussed. Furthermore, machine learning–based segmentation for medical image mining is depicted. Several related applications as well as challenges and future perspectives are also illustrated.

Chapter 12, "Biological Data Mining: Techniques and Applications," provides a comprehensive coverage of data mining for the concepts and applications of biological sequences. It includes related work of biological data mining applications with both fundamental concepts and innovative methods. Significant insights and suggested future research areas for biological data mining are introduced. This chapter is useful for the extraction of biological and clinical data ranging from genomic and protein sequences to DNA microarrays, protein interactions, biomedical images, and disease pathways.

Chapter 13, "Video Text Extraction and Mining," discusses the extraction of text information from videos and multimodal mining. This chapter provides a brief overview and classification of the methods used to extract text from videos and discusses their performances, their merits and drawbacks, available databases, their vulnerabilities, challenges, and recommendations for future development.

Chapter 14, "Deep Learning for Multimedia Content Analysis," discusses the principles and motivations regarding deep learning algorithms, such as deep belief networks, restricted Boltzmann machines, and the conventional deep neural network. It discusses the adaptation of deep learning methods to multimedia content analysis, ranging from low-level data such as audios and images to high-level semantic data such as natural language. The challenges and future directions are also addressed in this chapter.

Chapter 15, "Video-Image-Text Content Mining," focuses on videos and images that contain text data and useful information for indexing, retrieval, automatic annotation, and structuring of images. The extraction of this information can be executed in several phases from a digital video. This chapter explains in detail different phases of text extraction and the approaches used in every phase. The phases are preprocessing and segmentation, detection, localization, tracking, extraction, and recognition, respectively. In addition, the chapter discusses several suitable techniques according to the video type and phase. Mechanically, when these techniques have been applied, the text in video sequences will be extracted to provide useful information about their contents.

## Conclusion

Mining multimedia documents depends mainly on the features extracted from multimedia content, which includes text, audio, image, and video data from different domains. Multimedia content plays a significant role in building several applications in many domains, such as business, medicine, education, and military.

The chapters constituting this book reveal considerably how multimedia content can offer consistent information and useful relationships that can improve the document mining quality by

1. Introducing techniques and approaches for mining multimedia documents
2. Focusing on the document content: text, images, video, and audio
3. Providing an insight into the open research problems related to multimedia document mining
4. Offering an easy comprehension of the various document contents
5. Helping scientists and practitioners in choosing the appropriate approach for their problems

It is hoped that the chapters selected for this book will help professionals and researchers in this area to understand and apply the existing methods and motivate them to develop new approaches.

# *Editors*

**Wahiba Ben Abdessalem Karaa** is an Associate professor in the Department of Computer and Information Science at the University of Tunis. She obtained her PhD from Paris 7 Jussieu, France. Her research interests include natural language processing, text mining, image mining, and data mining. She is a member of the editorial boards of several international journals and is the editor in chief of the *International Journal of Image Mining* (IJIM).

**Nilanjan Dey** is an assistant professor in the Department of Information Technology at Techno India College of Technology, Kolkata. He is the editor in chief of the *International Journal of Rough Sets and Data Analysis*, IGI Global; managing editor of the *International Journal of Image Mining*; regional editor (Asia) of the *International Journal of Intelligent Engineering Informatics* (IJIEI); and associate editor of the *International Journal of Service Science, Management, Engineering, and Technology*. His research interests include medical imaging, soft computing, data mining, machine learning, rough sets, mathematical modeling and computer simulation, and the modeling of biomedical systems.

# Contributors

**Mariem Abdouli**
National School of Computer Sciences
and
RIADI Laboratory
ENSI
Manouba University
Manouba, Tunisia

**Adjan Abosolaiman**
Department of Computers and Information Technology
University of Taif
Taif, Saudi Arabia

**Eya Ben Ahmed**
Higher Institute of Applied Science and Technology
University of Sousse
Sousse, Tunisia

**Eman Alkhammash**
College of Computers & Information Technology
Taif University
Taif, Saudi Arabia

**Amira S. Ashour**
Department of Electronics and Electrical Communications Engineering
Tanta University
Tanta, Egypt

**Surekha Borra**
Department of ECE
K.S. Institute of Technology
Bangalore, Karnataka, India

**Sayan Chakraborty**
Bengal College of Engineering and Technology
Durgapur, West Bengal, India

**Ines Chouat**
Higher Institute of Management of Tunis
University of Tunis
Tunis, Tunisia

**Kawther Dridi**
Department of Computer Science
High Institute of Management of Tunis
Tunis University
Tunis, Tunisia

**Henda Ben Ghezela**
National School of Computer Sciences
and
RIADI Laboratory
ENSI
Manouba University
Manouba, Tunisia

**Dac-Nhuong Le**
Lecturer at Faculty of Information Technology
Haiphong University
Haiphong, Vietnam

**Monia Mannai**
Department of Computer Science
High Institute of Management of Tunis
Tunis University
Tunis, Tunisia

and

RIADI Laboratory
ENSI
Manouba University
Manouba, Tunisia

**Maram Mahmoud A. Monshi**
College of Computers & Information Technology
Taif University
Taif, Saudi Arabia

**Gia Nhu Nguyen**
Vice Dean, Graduate School
Duy Tan University, Viet Nam

**Prasenjit Kumar Patra**
Department of Information Technology
BCET, Durgapur, India

**Suresh Chandra Satapathy**
Department of Computer Science and Engineering
Anil Neerukonda Institute of Technology and Sciences
Visakhapatnam, Andra Pradesh, India

**Safa Selmi**
High Institute of Management of Tunis
Tunis University
Tunis, Tunisia

**Sabrine Benzarti Somai**
High Institute of Management of Tunis
Tunis University
Tunis, Tunisia

and

RIADI Laboratory
ENSI
Manouba University
Manouba, Tunisia

**Wafa Tebourski**
High Institute of Management of Tunis
Tunis University
Tunis, Tunisia

and

RIADI Laboratory
ENSI
Manouba University
Manouba, Tunisia

# Section I

# Motivation and Problem Definition

# 1

# *Mining Multimedia Documents: An Overview*

**Sabrine Benzarti Somai, Wahiba Ben Abdessalem Karaa, and Henda Ben Ghezela**

**CONTENTS**



**ABSTRACT** This chapter focuses on real-world problems that could involve multimedia mining. It proposes a literature review of approaches dealing with multimedia documents, taking into account various features extracted from multimedia content. The difference between static and dynamic media is explained. The multimodal nature of multimedia data creates an essential need for information fusion for its segmentation analysis, indexing, and even retrieval. Therefore, we present some approaches based on data fusion, audio, and video processing.

**KEY WORDS:** *multimedia mining, CBIR, high level, low level, data fusion, audio and video processing.*

## 1.1 Introduction

The amount of available data has become a problem for scientists who are not only responsible for the storage and preserving of these data but also for retrieving, categorizing, and analyzing these in order to use them in appropriate ways.

The multimedia document represents a real challenge for researchers. It is a sophisticated and complex data for the reason that a single document could contain diverse and varied features.

Mining multimedia documents is a rich and important area since when we say multimedia we cannot ignore images because even video is a sequence of images. Image mining has seen much progress in image treatment and retrieval.

The main purpose of this work is to present the multimedia document mining domain. Section 1.2 presents the mining multimedia process. Section 1.3 presents the multimedia mining architecture. Section 1.4 focuses on models that are used in multimedia data mining. The image mining field and some existing related works is presented in Section 1.5. The combination of text and images, called data fusion, is explained in Section 1.6, and some approaches related to this field such as deep learning are also presented. We focus on audio mining techniques in Section 1.7 and present some research works. Section 1.8 presents video processing, and the chapter ends with a conclusion.

## 1.2 Multimedia Mining Process

Multimedia are the most used data nowadays; they are available and have become the success key of many types of research. As a result, various processes exist, so why definitions should be treated carefully to avoid confusion.

Multimedia mining is a science interested in discovering knowledge hidden in a huge volume of images collection or a multimedia database in general. It is used to facilitate grouping, classification, finding hidden relation, and so on [1].

Multimedia mining has developed in the last years. It began by mining text using structured text [2,3], followed by context of image (bags of words), image feature (low level: color, structure, etc.), image features combined with experts analysis (high level), data fusion combining more than one media (image and text), and so on. Topics of multimedia data mining are varied: context- or content-based retrieval, similarity search [4], dimensional or prediction analysis, classification, and mining associations in multimedia data [5,6].

The multimedia mining process is divided into several steps. *Multimedia data collection* is the first stage of the mining process. Then, the *preprocessing* phase mines significant features from raw data. This level includes data cleaning, transformation, normalization, feature extraction, and so on.

The third phase of the multimedia mining process is *Learning*. It could be in a direct way if informative categories can be recognized at the preprocessing stage. The whole process depends enormously on the nature of raw data and the difficulty of the studied field. The output of preprocessing is the training set. Specified training set, is a learning model which has to be carefully chosen to learn from it and make the multimedia model more constant [7].

## 1.3 Multimedia Data Mining Architecture

The multimedia data mining processes have mostly the same architecture, to achieve their purpose in an appropriate way. It is divided into the following mechanisms [7]:

1. *Input selection* consists of the selection of the multimedia database used in the mining process. It facilitates the locating of multimedia content, which is the selected data as a subset of studied fields or data to be used for data mining.
2. *Data processing* depends on the nature of data; for example, the spatiotemporal segmentation is moving objects in image sequences in the videos and it is useful for object segmentation.
3. *Feature extraction*, also called the preprocessing step, includes integrating data from diverse sources; making choices of characterization or encoding some data fields to be used as inputs to the pattern finding the step. This stage is vital because of the complexity of certain fields that could involve data at different levels, and the unstructured nature of multimedia records.
4. *Finding similar pattern* is the aim of the entire data mining process. Some methods of finding similar pattern contain association, clustering, classification, regression, time-series, analysis, and so on.
5. *Evaluation of results* helps to assess results in order to decide whether the previous stage must be reconsidered or not.

## 1.4 Multimedia Data Mining Models

Several models are used in multimedia data mining. Their usage depends on the nature of the analyzed data, and the mining process purpose: It could be classification, knowledge extraction, or other goals. Multimedia mining techniques could be categorized in four major domains: classification, association rules, clustering, and statistical modeling [7].

### 1.4.1 Classification

Classification and predictive analysis are well used for mining multimedia data in many fields, particularly in scientific analysis as in astronomy and geoscientific analysis.

Classification is a technique for multimedia data analysis; it constructs data into categories divided into a predefined class label for a better effective and efficient use. It creates a function that well organizes data item into one of the several predefined classes, by inputting a training dataset and constructing a model of the class attributes based on the rest of the attributes. Decision tree classification is an example of the conceptual model without loss of exactness. Decision tree classification is a significant data mining method reported to image data mining applications. Also, hidden Markov model (HMM) is used for classifying multimedia data such as images and video.

The image data are often in large volumes and need considerable processing power, for example, parallel and distributed processing. The image data mining classification and clustering are judiciously associated to image analysis and scientific data mining and, hence, many image analysis techniques [7].

### 1.4.2 Clustering

The purpose of cluster analysis is to divide the data objects into multiple groups or clusters. Cluster analysis combines all objects based on their groups. Clustering algorithms can be divided into several methods: hierarchical methods, density-based methods, grid-based methods, model-based methods, k-means algorithm, and graph-based model [8]. In multimedia mining, clustering technique can be applied to assemble similar images, objects, sounds, videos, and texts.

### 1.4.3 Association Rules

Association rule is one of the most significant data mining techniques that aids in discovering hidden relations between data items in massive databases. Two major types of associations exist in multimedia mining: association between image content and non-image content features [1]. Mining the frequently occurring patterns among different images is the equivalent of mining the repeated patterns in a set of transactions. Multirelational association rule mining is the solution to exhibit the multiple reports for the same image. Correspondingly, multiple-level association rule techniques are used in image classification.

### 1.4.4 Statistical Modeling

Statistical mining models have as final objective the regulation of the statistical validity of test parameters and testing hypothesis, assuming correlation studies, and converting and preparing data for further analysis. This model creates correlations between words and partitioned image regions to establish a simple co-occurrence model [9].

## 1.5 Multimedia Mining: Image Mining

Image mining is the perception of unusual patterns and extraction of implicit and useful data from images stored in an enormous database. In other words, image mining tries to make and find associations between different images from a lot of images contained in databases.

As we mentioned, image processing begins by context or description content analysis, which is the text accompanying images; it could be a simple text, a report written by experts as the case in medical images or a metadata to annotate images as a manual annotation. But this way presents many difficulties and disadvantages—not only is it not objective, but also it is an expensive and a slow process. Researchers try to automate this process; they implement approaches based on image features as color, shape, texture, spatial relationships, and so on.

We can divide approaches developed for images processing in low level and high level. The low-level image processing is based on visual features such as color and texture.

We can also find some approaches that combine some image processing, such as Gaussian filtering, ellipse fitting, edge detection, and histogram thresholding.

However, high-level image processing is based on digging deep to search robust visual features by adapting and combining some techniques of machine learning and data mining with experts' knowledge.

The high-level image processing is characterized by intervention experts of the studied domain as rules [10] in order to help and improve the mining phase. This preprocessing task is very tedious—not only is it based on expert interviews that complicate the process, because of the nature of speech expressed in natural language that is ambiguous and informal, but also the translation of these rules into pixels or interesting objects, as constraints in the images set to be detected automatically. As a solution, the expert knowledge is usually expressed by class labels placed in images from the training set.

Content-based image retrieval (CBIR) is one of the fundamental field of research. It presents a real defies lengthily studied by multimedia mining and retrieval community for decades [5,11]. A CBIR purpose is to look for images through analyzing their visual contents, and therefore image representation is the heart of this method.

### 1.5.1 Low-Level Image Processing

The first and the most-used techniques in earlier multimedia data mining systems are those based on low-level image processing. It uses directly image features like color [12–14], texture [15–17], shape [18,19], and structure [20].

Several image querying systems founded on low level have been developed, for example, PhotoBook [21], The QBIC System (Query by Image and Video Content) [22], Virage [23], VisualSeek [24,25], and CENTRIST [26].

Images have many features; the color is still the most relevant one. First, it is a feature that is instantly perceived by the human eye. Second, it is a sensitive and a weak feature that could be easily influenced by other features such as luminosity; it remains a simple concept to understand and to implement.

### 1.5.2 High-Level Image Processing

The results obtained by using low-level content are often satisfactory. Nonetheless, there are some cases that need human intervention and therefore, a high level was invented. Also, research efforts are needed to bridge the gap among the high-level semantics, which users are interested in, and the low level that presents the image content. Human interpretation is compulsory; it could guide features extraction, retrieval, and querying, and finally result in an assessment.

The merge between the low and high levels gives other types of level-based classifications. For instance, J.P. Eakins [27] classified image features into three levels, going from the highly concrete to the most abstract. The first is the primitive level—its features include color, texture, shape, or the spatial location of image elements, in others words, the low level.

The second is the local semantic level, with features derived from the primitive features. Examples of queries by local semantic features are objects of a given type, such as "finding pictures with towers" or querying about the combination of objects such as "finding pictures with sky and trees." This type of queries is suitable for scenery images.

Finally, the thematic level or global semantic level features describe the meanings or topics of images. It is based on all objects and their spatial relationships in the image. For this, experts need high-level reasoning to derive the global meaning of all objects in the scene and discover the topic of the image. Some approaches have been developed that use semantic features to retrieve images such as IRIS [28], but results are still far away from the ambition and the expectation of researchers.

### 1.5.3 Application Using Image Data Mining

As presented earlier, content-based image retrieval (CBIR) systems use visual features to index images. The indexing phase prepares images for the principal task, which is to retrieve similar images.

Existing systems differ essentially in both extracting visual features to index images and the way they are queried. Diverse methods are adapted; there are systems using the image as query input, others allow a description of a list of constraints in the form of ad hoc queries that are in a particular language or as input in a user-friendly interface.

These systems look for similarity between images in the database by comparing features defined as constraints or signature (vector of features) extracted from the query with the appropriate features' vectors. The system presented in Reference 29 gives a query language for the description of spatial relationships within images. The DISIMA project [30] provides a visual query language VisualMOQL that has a pertinent expressiveness to describe constraints for visual features, as well as semantic image content. A point and click interface gives the user the opportunity to compose a query without knowing the query language itself. QBIC [22] and C-BIRD [32] offer means to describe the content of images in templates such as grids in various scales.

The similarity measures utilized in CBIR systems depend upon the visual features extracted and are commonly based on color, shape, texture, presence of given objects, spatial relationships, and so on.

As already mentioned, the color similarity is the most used measure and it is generally based on the general color distribution as a global color histogram or detected colors defined on grids overlapping the image. On the other hand, the objects' colors are very sensitive to light and, using only simple color similarity measure can give very poor and wrong results in the context of variations in illumination.

C-BIRD [32] proposed a measure established on chromaticity to match colors regardless of illumination. The texture resemblance diverges considerably from one system to another. For example, QBIC uses Tomura texture features [22], whereas C-BIRD utilizes four edge orientations (0°, 45°, 90°, 135°) and edge density [32].

The shape similarity discriminates between geometrical shapes within the images and shapes of objects painted in the image. The latter needs transformations because of angle, scale, and so on. Mostly, shapes designated in the objects' annotation in the images are utilized.

A significant effort has been made on the spatial resemblance measure [29]. This measure takes into account the closeness and adjacency of objects in the image. On another hand, it is presumed that the objects should be segmented and identified. This task is actually complex, so objects are manually recognized, annotated, and associated with a centroid. Images with centroids to represent objects are called symbolic images.

In DISIMA project [30], objects like buildings, vehicle, people, and animals are manually recognized and related with attributes such as type, name, function, and so on. The object similarity existence is the most delicate measure. With symbolic images, the recognition of objects is easy even with scaling, rotation, and translation.

The system CBIR [31] recognizes an object by constructing a sequence of descriptors as color and texture, gathered by locality. The system uses the notion of "blobs,*" founding a "blob world."

C-BIRD [32] offers to search by an object as a model. The system retrieves images containing a given object regardless of its orientation, scaling, or position in the image. The system is based on a three-step approach to reducing the searching space without using an index for object models. The search begins by pointing the first retrieving images containing the colors, texture, and shape of the given object, and then it starts searching the object in different orientations in pyramidal overlapped windows, and combining the object's color and texture properties in close areas with their respective centroids [33].

The last decades are regarded as the multimedia documents explosion, this huge amount of data contain hidden knowledge that need to be treated and analyzed to discover and exploit it in an appropriate and efficient way. Finding and developing new approaches became a necessity. But the diverse types of images present a real dilemma for researchers, so relevant research issues employ diverse mining techniques depending on the kind of treated image.

There are various types of images; the most treated are scenery and medical images. Each has its own characteristics, but scenery images are relatively simpler to analyze than others. It covers limited types of objects such as sky, tree, building, mountain, water, and so on. Consequently, the analyzing task of image features such as color, texture, spatial location of image elements, and shape is easier than other types of images.

### 1.5.4 Application of Image Data Mining in the Medical Field

Medical images are treated by various systems; the preprocessing level could be even more tedious, especially when the accuracy and the pertinence of mining task have to be very high.

Medical image processing is the field that offers researchers the occasion to further practice in order to try to eradicate the semantic gap. The cooperation between experts from different domains—computer scientists, doctors, radiologists—makes the multimedia mining task more arduous and multifaceted. The more we have an opinion the more we cannot arrive at a single and unified judgment. The medical imaging domain is characterized by its overlapping disciplines, but also it demands an overwork in order to integrate several information sources, and there are not enough available training datasets. All the mentioned difficulties make the medical imaging area a tough and a challenging field, but it has its clinical benefits [34,35].

Many systems have been developed; we will present briefly some of the systems in the following.

A well-known categorization scheme for diagnostic images is the IRMA† code. It classifies the visual content in four dimensions: (i) image modality as x-ray, ultrasound, and so on; (ii) body orientation; (iii) body region; and, finally, (iv) biological system. IRMA classes might help by way of concepts to build semantic meaningful visual signatures [36].

Deselaers et al. [6] used two features types: global feature and local feature. They used global features to describe the entire visual image content by one feature vector. The local features define specific locality in the images. The visual feature extracted could be simply based on color, shape, texture, or a mixture of those. To execute their system, they compare 19 images features using multiple datasets, including IRMA dataset containing 10,000 medical images [36].

---

* A blob is an elliptical area representing a rough localized coherent region in color and textual space.
† Medical image categorization systems.

Iakovidis et al. get encouraging medical image retrieval results on the IRMA dataset. They generated visual signature by means of cluster wavelet coefficients (the wavelet transforms is a mathematical model well used to represent texture features [17]) and estimate the distributions of clusters by means of Gaussian mixture models with an expectation-maximization algorithm [37]. Quellec et al. adapted the wavelet basis 16 to optimize retrieval performance inside a given image collection [38]. Chatzichristofis et al. proposed a merged image descriptor locating brightness and texture characteristics for medical image retrieval [39].

Rahman et al. [40] proposed a CBIR framework exploiting class probabilities of several classifiers as visual signatures and cosine similarity for retrieval task. Class probabilities are estimated from binary support vector machine (SVM) classifiers. For diverse low-level visual feature, concepts values similarity are calculated distinctly and merged by linear combination scheme that optimizes corresponding weights for each query. The weight optimization includes automatic pertinence estimation centered on classifier synthesis over low-level feature spaces.

The framework was assessed on the Image CLEF 2006 medical dataset using 116 IRMA categories and four low-level visual features (MPEG-7 Edge Histogram and Color Layout, GLCM-based texture features, and block-based gray values). In 2011, the authors proposed an ameliorated retrieval scheme based on similar approaches [41].

Güld et al. [42] presented a generic framework dedicated to medical image retrieval systems developed by the IRMA project [36]. The proposed framework enables flexible and effective development and deployment of retrieval algorithms in a distributed environment with web-based user interfaces.*

Zhou et al. proposed a framework for semantic CBIR medical images retrieval. They highlighted the necessity of a scalable semantic retrieval system. Their system is flexible; it is well adaptable to different image modalities and anatomical regions. It could incorporate external knowledge [31]. The architecture integrates both symbolic and subsymbolic image feature content extraction and proposes a semantic reasoning. To implement their system, they described a semantic anatomy tagging engine called ALPHA, using a new approach dedicated to deformable image segmentation through combining hierarchical shape decomposition, and CBIR.

LIRE† is a Java library supporting content-based text and image retrieval [39,43]. It affords a list of diverse global and local image feature extractors and efficient indexing techniques for images and text based on Lucene.‡ Mammography is well exploited to detect cancer; however, it needs major preprocessing before use. Images have to be treated to highlight interesting zones such as noise elimination; dealing with the dark background or over-brightness. An automatic retinal photography classification system was developed to discover retinopathy (a common cause of blindness among diabetic patients). The system aim is image analysis in order to recognize optic disc anomalies, tortuous blood vessels, or abnormal lesions (exudates). The challenging task is to extract the visual features that illustrate the optic disc, the vessels, or the exudates. The system combines image processing, like ellipse fitting, edge detection, histogram thresholding, Gaussian filtering, and machine learning techniques such as Bayesian classifiers.

Another system proposed in Reference 44 uses association rule mining to classify retinal photography into groups normal and abnormal, using features (blood vessels, patches,

* http://irma-project.org/onlinedemos.php.
† http://www.semanticmetadata.net/lire/.
‡ http://lucene.apache.org/.

optic disc) wisely extracted from the images after several image processing. The experimented system had an accuracy of 88%, detecting abnormal retinas on real datasets.

The Queensland University project classifies objects in images in order to detect early signs of cancer of the cervix by detecting abnormal cells in pap smear slides [45]. The system analyzes thousands of cells per patient to perceive cells that do not need checking with the aim of saving time to human operators. An original technique for segmenting the cell nucleus was developed using hidden Markov model to classify the cells into two clusters, easy observation and hard observation, realizing more than 99% accuracy.

An innovative method for fast detection of areas containing doubtful restricted lesions in mammograms is presented. The method locates the interesting regions in the image using a radial-basis-function neural network after it differentiates between the normal and the abnormal mammograms using regular criteria based on statistical features. To localize areas of interest in the image, the system used a neural network.

The system presented in Reference 46 uses association rules to sort mammograms centered on the type of tumor. The used features in the item sets are descriptive attributes from the patient record and the radiologist tumor annotation with extracted visual features from the mammogram. The primary results seem encouraging but nonconclusive.

The biclustering is well used for image segmentation for detecting interesting zone to locate tumors and affected organs by cancer [47].

There are semantic images researchers based on ontologies. In this purpose, we present the semantic search approach using polyps' endoscopic images. This research is based on a standard reasoning adequacy logic description associated with the ontology of polyps and a suitable image annotation mechanism [48].

## 1.6 Text and Image Feature Retrieval: Data Fusion

The multimedia mining domain is up, it usually pursuits data and user need progression. It starts by text retrieval, then images retrieval, video retrieval, and so on. Nowadays, data types are overlapped; we cannot distinguish or separate heterogeneous data. Hence, the multimedia mining techniques should be up to date and treat mixed information; data fusion, also called metadata, is the consequence of this phenomena. Merging text and visual retrieval leads to the most general problem of data fusion [49]. The main idea is to combine many information sources to increase retrieval efficiency and pertinence.

Caicedo et al. presented a method for detecting relevant images for the query topic by combining visual features and text data using latent semantic kernels by adding image kernel and text kernel functions together [50].

Moulin [51] the main purpose is the representation of multimedia documents as a model that allows exploiting the documents, combining text and images for classification or information retrieval systems. Moulin et al. adapted a new feature to limit the vocabulary (CCDE) and proposed a new method to solve the problem of multilabel (MCut). To represent images they used a model based on visual words bags weighted tf-idf. Moulin et al. assessed their work on conventional image collections CLEF and INEX mining. The limit of this approach is the fact of considering just flat text regardless its structure.

Bassil proposed a hybrid information retrieval model dedicated to web images. The approach is based on color base image retrieval (color histogram) and keyword information retrieval technique for embedded textual metadata (HTML). Term weighting is based on a novel measure VTF-IDF (variable term frequency-inverse document frequency). The author used variable to design terms, respecting not only the HTML tag structure but also its location where tags appears [52].

There are many researchers trying to study the impact of structures of multimedia documents on retrieval task. There are works representing the points of interest of an image in the form of a graph. To compare two images, it is equivalent to compare the graphs that represent each one [3].

Motivated by recent successes of deep learning techniques for computer vision and other applications, Cheng developed a learning approach [53] to recognize the three graphics types: graph, flowchart, and diagram. He used a data fusion approach to combine information from both text and image sources. He developed *method applied*: a hybrid of an evolutionary algorithm (EA) and binary particle swarm optimization (BPSO) to find an optimum subset of extracted image features. To select the optimal subset of extracted text features, he used Chi-square statistic and information gain metric, which along with image features are input to multilayer perceptron neural network classifiers, whose outputs are characterized as fuzzy sets to determine the final classification result. To evaluate the performance of their approach, he used 1707 figure images extracted from a test subset of *BioMedCentral* journals extracted from U.S. National Library of Medicine's PubMed Central repository giving 96.1% classification accuracy [53].

Also, Beibei Cheng explored a framework of deep learning with application to CBIR tasks with an extensive set of experimental studies by examining a state-of-the-art deep learning method (convolutional neural networks: CNNs) for CBIR tasks under varied settings. To implement the CNNs learning, they used the similar framework as discussed in Reference 54 by adjusting their accessible released C++ implementation. This approach is executed on the "ILSVRC-2012"* dataset from ImageNet and found state-of-the-art performance with 1000 categories and more than one million training images [53].

## 1.7 Audio Mining

Audio mining has a primordial role in multimedia applications; the audio data contain sound, MP3 songs, speech, music, and so on.

Audio data mining gathers diverse techniques in order to search, analyze, and route with wavelet transformation the audio signal content.

The audio processing could use band energy, zero crossing rate, frequency centroid, pitch period, and bandwidth as input features for the mining process [55].

Audio data mining is widely used in automatic speech recognition, which analyzes the signal in order to find any speech within the audio.

Many types of research are done and many applications are developed related to the audio mining field based on the extraction and characterization of audio features. Radhakrishnan et al. [56] proposed a content adaptive representation framework for event discovery based on audio features from "unscripted" multimedia like surveillance data and sports. Radhakrishnan et al. used the hypothesis that interesting events happen rarely in a background of uninteresting events, the audio sequence is considered as a time series, and temporal segmentation is achieved to identify subsequences that are outliers constructed on a statistical model of the series.

* http://www.image-net.org/challenges/LSVRC/2012/.

Chu et al. [57] modulated the statistical characteristics of audio events as a hierarchical method over a time series to achieve semantic context detection. Specifically, modeling at the two separate levels of audio events and semantic context is proposed to bridge the gap between low-level audio features and semantic concepts.

Czyzewski [58] used knowledge data discover (KDD) methods to analyze audio data and remove noise from old recordings.

## 1.8 Video Mining

The aim of video mining is to find the interesting patterns from a large amount of video data. The processing phase could be indexing, automatic segmentation, content-based retrieval, classification, and detecting triggers.

Zhang and Chen [59] presented a new approach to extract objects from video sequences, which is based on spatiotemporal independent component analysis and multiscale analysis. The spatiotemporal independent component analysis is the first step executed to recognize a set of preliminary source images, which contain moving objects. The next phase is using wavelet-based multiscale analysis to increase the accuracy of video object extraction.

Liu et al. [60] proposed a new approach for performing semantic analysis and annotation of basketball video. The model is based on the extraction and analysis of multimodal features, which include visual, motion, and audio information. These features are first combined to form a low-level representation of the video sequence. Based on this representation, they then utilized domain information to detect interesting events, such as when a player performs a successful shot at the basket or when a penalty is imposed for a rule violation, in the basketball video.

Hesseler and Eickeler [61] proposed a set of algorithms for extracting metadata from video sequences in the MPEG-2 compressed domain. The principle is the extracted motion vector field; these algorithms can deduce the correct camera motion, which permit motion recognition in a limited region of interest for the aim of object tracking, and perform cut detection.

Fonseca and Nesvadba [62] introduced a new technique for face detection and tracking in the compressed domain. More precisely, face detection is performed using DCT coefficients only, and motion information is extracted based on the forward and backward motion vectors. The low computational requirement of the proposed technique facilitates its adoption on mobile platforms.

## 1.9 Conclusion

The multimedia data mining field is promising because it covers almost every domain. However, it needs laborious and tedious work since it covers several and overlapping data and areas [63].

Furthermore, the specificity of multimedia data, which need extra treatment and could be ambiguous, makes researcher task increasingly more challenging.

The preprocessing phase, which launches the multimedia mining procedure, is the most vital and thoughtful phase of the knowledge discovery process. Mainly, preprocessing can "make-it or break-it."

Preprocessing multimedia data before mining and searching process concerns extracting or underlining some visual features in the data that may well be relevant in the mining task.

Often in multimedia mining, and image mining especially, we speak about high level, because the choice of features is determined by interviewing domain experts to capture their knowledge as a set of semantic features and rules. These high-level features and rules are later converted into pixel-level constraints and automatically extracted from the images. This process, conversely, is not usually probable as the expressiveness of rules or descriptions given by experts is not always exact, clear, and precise enough to be turned into pixel-level constraints for various domains or basically other new images.

Image or video treatment is an entire range of various image-processing techniques to identify and extract key visual features from the images, comparable to precarious medical symptoms in the case of medical images. The main defy with mining medical images is to come up with worthy image models and have a relevant process for diverse domain issues by identifying and extracting the right visual features.

An additional common concern is the similarity matching concept obvious for image mining. These challenges are strongly associated with compound object recognition and image understanding, difficulties that are addressed by computer vision and artificial intelligence research communities. Recent researches are concentrated on the perception of deep learning, which gives very encouraging and promising results [53,64].

## References


1. Manjunath, T. N., Hegadi, R. S., and Ravikumar, G. K. (2010). A survey on multimedia data mining and its relevance today. *IJCSNS, 10*(11), 165–170.
2. Idarrou, A. (2013). Entreposage de documents multimédias: comparaison de structures. (Doctoral dissertation), Toulouse 1, Toulouse, France.
3. Torjmen, M. (2009). Approches de recherchemultimédiadans des documents semi-structurés: utilisation du contextetextueletstructurel pour la sélectiond'objetsmultimédia. (Doctoral dissertation), Université de Toulouse, Université Toulouse III-Paul Sabatier, Toulouse, France.
4. Arevalillo-Herráez, M. and Ferri, F. J. (August 2010). Interactive image retrieval using smoothed nearest neighbor estimates. In *Joint IAPR International Workshops on Statistical Techniques in Pattern Recognition (SPR) and Structural and Syntactic Pattern Recognition (SSPR)* (pp. 708–717). Springer, Berlin, Germany.
5. Lew, M. S., Sebe, N., Djeraba, C., and Jain, R. (2006). Content-based multimedia information retrieval: State of the art and challenges. *ACM Transactions on Multimedia Computing, Communications, and Applications (TOMM)*, *2*(1), 1–19.
6. Deselaers, T., Keysers, D., and Ney, H. (2008). Features for image retrieval: An experimental comparison. *Information Retrieval, 11*(2), 77–107.
7. Vijayarani, S. and Sakila, A. (2015). Multimedia mining research—an overview. *International Journal of Computer Graphics & Animation*, *5*(1), 69.
8. Manjunath, R. and Balaji, S. (2014). Review and analysis of multimedia data mining tasks and models. *International Journal of Innovative Research in Computer and Communication Engineering, 2*, 124–130.
9. Jiawei, H. and Kamber, M. (2001). *Data Mining: Concepts and Techniques*, vol. 5. Morgan Kaufmann, San Francisco, CA.

10. Burl, M. C., Fowlkes, C., and Roden, J. (1999). Mining for image content. In *Systemics, Cybernetics, and Informatics/Information Systems: Analysis and Synthesis,* Orlando, FL, July 1999.
11. Forsyth, D. A., Malik, J., Fleck, M. M., Greenspan, H., Leung, T., Belongie, S., Carson, C. et al. (April 1996). Finding pictures of objects in large collections of images. In *International Workshop on Object Representation in Computer Vision* (pp. 335–360). Springer, Berlin, Germany.
12. Swain, M. J. and Ballard, D. H. (1991). Color indexing. *International Journal of Computer Vision, 7*(1), 11–32.
13. Pass, G., Zabih, R., and Miller, J. (1996). Comparing images using color coherence vectors. In *Proceedings of ACM Multimedia*, vol. 96 (pp. 65–73). Boston, MA.
14. Mokhtarian, F., Abbasi, S., and Kittler, J. (September 1996). Robust and E cient shape indexing through curvature scale space. In *Proceedings of the 1996 British Machine and Vision Conference BMVC*, vol. 96.
15. Manjunath, B. S. and Ma, W. Y. (1996). Texture features for browsing and retrieval of image data. *IEEE Transactions on Pattern Analysis and Machine Intelligence, 18*(8), 837–842.
16. Dougherty, E. R. and Pelz, J. B. (1989). Texture-based segmentation by morphological granulometrics. In *Advanced Printing of Paper Summaries, Electronic Imaging*, 89, 408–414.
17. Do, M. N. and Vetterli, M. (2002). Wavelet-based texture retrieval using generalized Gaussian density and Kullback-Leiblerdistance. *IEEE Transactions on Image Processing, 11*(2), 146–158.
18. Pass, G., Zabih, R., and Miller, J. (February 1997). Comparing images using color coherence vectors. In *Proceedings of the Fourth ACM International Conference on Multimedia*, Boston, MA, November 1996 (pp. 65–73). ACM.
19. Jain, A. K. and Vailaya, A. (1996). Image retrieval using color and shape. *Pattern Recognition, 29*(8), 1233–1244.
20. Ahuja, N. and Rosenfeld, A. (1981). Mosaic models for textures. *IEEE Transactions on Pattern Analysis and Machine Intelligence, 3*(1), 1–11.
21. Pentland, A. P., Picard, R. W., and Scarloff, S. (April 1994). Photobook: Tools for content-based manipulation of image databases. In *IS&T/SPIE 1994 International Symposium on Electronic Imaging: Science and Technology* (pp. 34–47).
22. Flickner, M., Sawhney, H., Niblack, W., Ashley, J., Huang, Q., Dom, B., Gorkani, M. et al. (1995). Query by image and video content: The QBIC system. *Computer, 28*(9), 23–32.
23. Bach, J. R., Fuller, C., Gupta, A., Hampapur, A., Horowitz, B., Humphrey, R., Jain, R., and Shu, C.-F. (1996). The Virage image search engine: An open framework for image management. In: I. K. Sethi, R. C. Jain (eds.), *Proceedings of the SPIE Conference on Storage & Retrieval for Image and Video Databases IV*, vol. 2670, San Jose, CA (pp. 76–87).
24. Smith, J. R. and Chang, S. F. (February 1997). VisualSEEk: A fully automated content-based image query system. In *Proceedings of the Fourth ACM International Conference on Multimedia*, Boston, MA, November 1996 (pp. 87–98). ACM.
25. Lehmann, T. M., Gold, M. O., Thies, C., Fischer, B., Spitzer, K., Keysers, D., and Ney, H. (2004). Content-based image retrieval in medical applications. *Methods of Information in Medicine, 43*(4), 354–361.
26. Wu, J. and Rehg, J. M. (2011). CENTRIST: A visual descriptor for scene categorization. *IEEE Transactions on Pattern Analysis and Machine Intelligence, 33*(8), 1489–1501.
27. Eakins, J. P. (2002). Towards intelligent image retrieval. *Pattern Recognition, 35*(1), 3–14.
28. Alshuth, P., Hermes, T., Klauck, C., Kreyß, J., and Röper, M. (1996). Iris-image retrieval for images and videos. In *Proceedings of First International Workshop of Image Databases and MultiMedia Search*, IDB-MMS, August 1996 (pp. 170–178).
29. Sistla, A. P., Yu, C., Liu, C., and Liu, K. (September 1995). Similarity-based retrieval of pictures using indices on spatial relationships. In *VLDB* (pp. 619–629).
30. Oria, V., Ozsu, M. T., Xu, B., Cheng, I., and Iglinski, P. J. (July 1999). VisualMOQL: The DISIMA visual query language. In *IEEE International Conference on Multimedia Computing and Systems*, Italy, 1999, vol. 1 (pp. 536–542). IEEE.

31. Zhou, X. S., Zillner, S., Moeller, M., Sintek, M., Zhan, Y., Krishnan, A., and Gupta, A. (July 2008). Semantics and CBIR: A medical imaging perspective. In *Proceedings of the 2008 International Conference on Content-Based Image and Video Retrieval*, Niagara Falls, Ontario, Canada, July 7–9, 2008 (pp. 571–580). ACM.
32. Li, Z. N., Zaïane, O. R., and Yan, B. (August 1998). C-BIRD: Content-based image retrieval from digital libraries using illumination invariance and recognition kernel. In *Proceedings of Ninth International Workshop on Database and Expert Systems Applications 1998* (pp. 361–366). IEEE.
33. Zaıane, O. R. (1999). Resource and knowledge discovery from the internet and multimedia repositories. Doctoral dissertation, Simon Fraser University, Burnaby, British Columbia, Canada.
34. Oliva, A. and Torralba, A. (2001). Modeling the shape of the scene: A holistic representation of the spatial envelope. *International Journal of Computer Vision*, *42*(3), 145–175.
35. Müller, H., Michoux, N., Bandon, D., and Geissbuhler, A. (2004). A review of content-based image retrieval systems in medical applications—Clinical benefits and future directions. *International Journal of Medical Informatics*, *73*(1), 1–23.
36. Lehmann, T. M., Schubert, H., Keysers, D., Kohnen, M., and Wein, B. B. (May 2003b). The IRMA code for unique classification of medical images. In *Proceedings SPIE*, vol. 5033 (pp. 440–451). International Society for Optics and Photonics.
37. Iakovidis, D. K., Pelekis, N., Kotsifakos, E.E., Kopanakis, I., Karanikas, H., and Theodoridis, Y. (2009). A pattern similarity scheme for medical image retrieval. *IEEE Transactions on Information Technology in Biomedicine*, 13, 442–450.
38. Quellec, G., Lamard, M., Cazuguel, G., Cochener, B., and Roux, C. (2010). Wavelet optimization for content-based image retrieval in medical databases. *Medical Image Analysis*, *14*(2), 227–241.
39. Lux, M. and Chatzichristofis, S. A. (October 2008). Lire: Lucene image retrieval—An extensible java cbir library. In *Proceedings of the 16th ACM International Conference on Multimedia*, Vancouver, British Columbia, Canada, October 2008 (pp. 1085–1088). ACM.
40. Rahman, M. M., Desai, B. C., and Bhattacharya, P. (2008). Medical image retrieval with probabilistic multi-class support vector machine classifiers and adaptive similarity fusion. *Computerized Medical Imaging and Graphics*, *32*(2), 95–108.
41. Rahman, M. M., Antani, S. K., and Thoma, G. R. (2011). A learning-based similarity fusion and filtering approach for biomedical image retrieval using SVM classification and relevance feedback. *IEEE Transactions on Information Technology in Biomedicine*, *15*(4), 640–646.
42. Güld, M. O., Thies, C., Fischer, B., and Lehmann, T. M. (2007). A generic concept for the implementation of medical image retrieval systems. *International Journal of Medical Informatics*, *76*(2), 252–259.
43. Lux, M. and Marques, O. (2013). Visual information retrieval using java and lire. *Synthesis Lectures on Information Concepts, Retrieval, and Services*, *5*(1), 1–112.
44. Hsu, W., Lee, M. L., Liu, B., and Ling, T. W. (August 2000). Exploration mining in diabetic patients databases: Findings and conclusions. In *Proceedings of the Sixth ACM SIGKDD International Conference on Knowledge Discovery and Data Mining*, Boston, MA (pp. 430–436). ACM.
45. Bamford, P. and Lovell, B. (2001). Method for accurate unsupervised cell nucleus segmentation. In *Proceedings of the 23rd Annual International Conference of the IEEE Engineering in Medicine and Biology Society*, Istanbul, Turkey, 2001, vol. 3 (pp. 2704–2708). IEEE.
46. Antonie, M. L., Zaiane, O. R., and Coman, A. (2001). Application of data mining techniques for medical image classification. In *MDM/KDD 2001* (pp. 94–101).
47. Sayana, S. and Pratheba, M. (2014). Detection of cancer using biclustering. *International Journal of Innovative Research in Computer and Communication Engineering*, 2(SI 1), 2409–2415.
48. Chabane, Y. and Rey, C. Annotation et recherchesémantiqued'images en gastroentérologie. SIIM 2013, 2e édition du Symposium sur l'Ingénierie de l'Information Médicale SIIM 2013, Lille, 1 Juillet 2013.
49. Valet, L., Mauris, G., and Bolon, P. (July 2000). A statistical overview of recent literature in information fusion. In *Proceedings of the Third International Conference on Information Fusion*, Stockholm, Sweden, *2000 (FUSION 2000)*, vol. 1 (pp. MOC3–MOC22). IEEE.

50. Caicedo, J. C., Moreno, J. G., Niño, E. A., and González, F. A. (March 2010). Combining visual features and text data for medical image retrieval using latent semantic kernels. In *Proceedings of the International Conference on Multimedia Information Retrieval*, Philadelphia, PA (pp. 359–366). ACM.
51. Moulin, C. (2011). Modélisation de documents combinanttexteet image: Application à la catégorisation et à la recherched'informationmultimédia. Doctoral dissertation, Université Jean Monnet, Saint Etienne, France.
52. Bassil, Y. (2012). Hybrid information retrieval model for web images. *arXiv preprint arXiv:1204.0182*.
53. Cheng, B., Stanley, R. J., Antani, S., and Thoma, G. R. (August 2013). Graphical figure classification using data fusion for integrating text and image features. In *12th International Conference on Document Analysis and Recognition* (pp. 693–697). IEEE.
54. Krizhevsky, A., Sutskever, I., and Hinton, G. E. (2012). Imagenet classification with deep convolutional neural networks. In *Advances in Neural Information Processing Systems* Inc., MIT Press (pp. 1097–1105).
55. More, S. and Mishra, D. K. (2012). Multimedia data mining: A survey. *Pratibha: International Journal of Science, Spirituality, Business and Technology (IJSSBT)*, *1*(1).
56. Radhakrishnan, R., Divakaran, A., and Xiong, Z. (October 2004). A time series clustering based framework for multimedia mining and summarization using audio features. In *Proceedings of the Sixth ACM SIGMM International Workshop on Multimedia Information Retrieval*, New York, October 15–16, 2004 (pp. 157–164). ACM.
57. Chu, W. T., Cheng, W. H., and Wu, J. L. (2006). Semantic context detection using audio event fusion: Camera-ready version. *EURASIP Journal on Applied Signal Processing*, *2006*, 181.
58. Czyzewski, A. (December 1996). Mining knowledge in noisy audio data. In *KDD*, Portland, OR (pp. 220–225).
59. Chen, X. and Zhang, C. (December 2006). An interactive semantic video mining and retrieval platform—Application in transportation surveillance video for incident detection. In *Sixth International Conference on Data Mining (ICDM'06)* (pp. 129–138). IEEE.
60. Liu, S., Xu, M., Yi, H., Chia, L. T., and Rajan, D. (2006). Multimodal semantic analysis and annotation for basketball video. *EURASIP Journal on Advances in Signal Processing*, *2006*(1), 1–13.
61. Hesseler, W. and Eickeler, S. (2006). MPEG-2 compressed-domain algorithms for video analysis. *EURASIP Journal on Applied Signal Processing*, *2006*, 186.
62. Fonseca, P. M. and Nesvadba, J. (2006). Face tracking in the compressed domain. *EURASIP Journal on Applied Signal Processing*, *2006*, 187.
63. Guan, L., Horace, H. S. Ip, Lewis, P. H., Wong, H. S., and Muneesawang, P. (2005). Information mining from multimedia databases. *Journal on Applied Signal Processing*, Hindawi Publishing Corporation EURASIP(2006), Article ID 49073, 1–3.
64. Singh, A. V. (2015). *Content-Based Image Retrieval Using Deep Learning*, thesis, Rochester Institute of Technology, New York.

# Section II

# Text Mining Using NLP Techniques

# 2

# *Fuzzy Logic for Text Document Clustering*

**Kawther Dridi, Wahiba Ben Abdessalem Karaa, and Eman Alkhammash**

**CONTENTS**



**ABSTRACT** The difference between computers and the human brain is explained by reasoning. This means that the human brain can use uncertain data, but computers reason with precise data [1]. Nowadays, fuzzy logic has become an important solution to reduce the difference between the human brain and computers.

Fuzzy logic has become an important field of study, thanks to its ability to help researchers to manipulate data that were not accurate and not precise; it can manipulate vague propositions. But classical logic deals with exact values of variables, which means it only supports precise data; however, it cannot handle uncertain and imprecise data. In our work, we propose an approach based on fuzzy logic and Euclidean distance metric for text document clustering. The idea is to search for the similarities and dissimilarities between biological documents to facilitate classification.

**KEY WORDS:** *fuzzy logic, text document clustering, similarity, distance measurements.*

## 2.1 Introduction

Clustering is an important task of assigning objects to clusters or groups such that a collection of objects in the same class are similar and objects from different classes are dissimilar. Examples of clustering tools are C4.5 [2], k-means clustering algorithm [3], and self-organizing map (SOM) [4].

In classical logic, each object has variables with accurate and precise values. Thus, each object belongs to a single class. However, information is not always precise and complete and can be uncertain and inaccurate. So each person can find themselves in different situations where they use incomplete information and reason with them to make decisions. In such cases, fuzzy logic can help to solve the problem [5].

Due to the large number of documents available on the web, retrieval of such documents is a difficult task. A variety of classification algorithms have been developed to resolve this problem. In classification problems, we are often confronted with overlapping classes, which have limitless boundaries and cannot be presented in a specific partition.

Nowadays, getting the right information at the right time has become a necessity. However, upon classification of a text document, a classifier may not be able to determine the class associated with the document. The classifier may also confuse the terms of the document with another one, which may lead to a rejection. So, to help prevent this issue, we propose to use an approach that is based on fuzzy logic.

In fact, fuzzy logic has been recently presented as a model distinguished by its ability to handle uncertain and imprecise information. It was introduced as an extension of classical logic.

In our work, we propose a new approach that facilitates searching and classification tasks. Our approach combines fuzzy logic and Euclidean distance metric for text document clustering. The key idea is to search for the similarities and dissimilarities between documents to facilitate classification.

This chapter is organized as follows. We provide an overview of fuzzy logic in Section 2.2. In Section 2.3, we present the proposed algorithm for text document clustering. In Section 2.4, we evaluate the performance of our proposed algorithm in text document clustering, and finally, we give the conclusion in Section 2.5.

## 2.2 Background

### 2.2.1 Fuzzy Logic

Fuzzy logic is presented as a generalization of classical logic. It was introduced by Zadeh [5] to resolve the problems related to the imprecision and incompleteness of information. Fuzzy logic includes probability theory and also other theories such as Dempster–Shafer theory (or evidence theory), possibility theory, and so on.

In fuzzy logic, there are no two alternatives or a whole continuum of truth values for logical propositions. A proposition A can have the truth value 0.4, and its complement can have the truth value 0.5. According to the type of negation operator that is used, the truth values must not be added upto 1. First, fuzzy logic can be applied as an interpretation model for the properties of neural networks by giving more description of their performance. Second, it can be used to specify networks without having to apply a learning algorithm.

Unlike Boolean sets, where the characteristic function takes only two values, either 0 or 1, the function of the fuzzy logic can take current values between 0 and 1. Therefore, fuzzy logic considers the notion of belonging of an element to a set of classes as a function that can take values from 0 to 1.

So, the difference between the classical sets and the fuzzy sets is presented by introducing a membership function.

Fuzzy sets were introduced by Zadeh to represent and manipulate data that were not precise but rather fuzzy.

A fuzzy set E is characterized by a membership function $U_E$ that associates to each object in the universe, its membership degree $U_E(x)$, in the interval [0, 1]. Note that X is a collection of objects. X is called the universe of discourse, and x represents an element of X.

The more $U_E(x)$ tends to 1, the more x belongs to E. Where $U_E(x)$ takes only the values 0 and 1, fuzzy set E is a conventional subset of X. A fuzzy set is constituted by a set of the fuzzy subset.

For example, let us consider three fuzzy sets—"young," "mature," and "old"—that will be defined by a membership function $U_E$ associated with each person in the universe X. Note that X is a collection of persons, $X = \{x_1, x_2, x_3, \ldots, x_n\}$, where the degree of the membership function is expressed by a real number in the interval [0, 1]. So, this degree can be interpreted as a continuous process in which the membership of a person to the set of adults takes values between 0 and 1.

There are many other examples other than the example of age, such as that of fast and slow. The concepts of mature, old, and young or of the adjectives fast and slow are imprecise, incomplete, and inaccurate but easy to interpret in a given context. This is what the fuzzy logic tries to accomplish.

### *2.2.1.1 Fuzzy Operators*

In the case of classical logic, the operators used are AND, OR, and NOT to express the intersection, union, and complement operations.

But in fuzzy logic, the used operators need to be defined as functions for all possible fuzzy values from 0 to 1 [5,6].

The generalization of these operators is presented as follows [5,6]:

- The union of two fuzzy sets with membership functions is presented as the maximum.
- The intersection of two fuzzy sets with membership functions is presented as the minimum.
- The complement of two fuzzy sets with membership functions is defined as the complement.

So the union, intersection, and complement are presented [5,6] as follows:

*Union*: The membership of the union of bivalent fuzzy sets A and B is

$$\mu A \cup B(x) = \max(\mu A(x), \mu B(x)) \forall x \in X$$

Interpretation functions: T-conorms

*Intersection*: The membership of the intersection of fuzzy sets A and B can be defined as

$$\mu A \cap B(x) = \min(\mu A(x), \mu B(x)) \forall x \in X$$

Interpretation functions: T-norms

*Complement*: The complement of membership functions can be defined as

$$\mu A^c(x) = 1 - \mu A(x)\ \forall x \in X$$

Interpretation functions: S-norms

#### 2.2.1.2 Membership Function

The membership function is used to measure the membership degree of every element to its group.

It can take current values between 0 and 1. Therefore, fuzzy logic considers the notion of belonging of an element to a set of classes as a function that can take values from 0 to 1.

#### 2.2.1.3 Fuzzy Logic and Application Fields

##### 2.2.1.3.1 *Fuzzy Logic in Machine Learning and Data Mining*

There are several approaches to extract models or patterns from data. These approaches have received considerable attention in the fuzzy set community, which is the reason why Hüllermeier [7] chose to present some contributions that fuzzy set theory can make in many fields, such as machine learning and data mining.

To adapt to the progress in data acquisition and storage technology, knowledge discovery in databases (KDD) has emerged as a new discipline. It refers to the process of identifying structure in data. In other words, KDD process is data mining.

Data mining puts special emphasis on the analysis of data sets in terms of scalability and efficiency. The goal in data mining is not to discover global models but rather to discover local patterns of interest. It is of an explanatory nature, and models discovered in data sets are of a descriptive, not a predictive, nature. There are several steps that characterize the data mining process. These steps are data cleaning, data integration, data selection, data transformation, data mining, evaluation of patterns, and finally knowledge presentation.

Machine learning focuses on prediction, based on known properties learned from the training data. It is based on the construction of systems that can learn from data.

So the contribution that fuzzy set theory can make in the field of data mining is expressed by the development of tools and technologies that have the potential to support all the steps of the data mining process. In particular, fuzzy set theory can be employed especially in two phases: data selection and data preparation. It is used for modeling vague data in terms of fuzzy set, creating summaries of fuzzy data, and so on.

In addition, fuzzy set theory can produce patterns that are more comprehensible and robust. It can also contribute to the modeling and processing of multiple forms of incomplete information [7].

##### 2.2.1.3.2 *Fuzzy Logic Used for the Sendai Subway System*

The areas of potential fuzzy implementation are numerous, including control areas. Fuzzy logic can control or evaluate systems by using rules that refer to indefinite quantities. Fuzzy systems often define their rules from experts. If no expert defines the rules, adaptive fuzzy systems learn the rules by observing how people manipulate real systems. One of the most applications, namely the Sendai Subway system in Sendai, Japan, was recently addressed using fuzzy logic. The Nanboku line, developed by Hitachi, employed a fuzzy controller to direct the train. Researchers used fuzzy rules to accelerate, slow, and brake the subway trains more smoothly. This is also an example of the earlier acceptance of fuzzy logic in the East [1].

#### 2.2.1.3.3 *Fuzzy Logic Used for Air-Conditioning Systems*

These systems use fuzzy logic to control the heating and cooling, which saves energy by making the system more efficient. Therefore, fuzzy logic represents the notion of belonging of an element to a set of classes as a function that can take values from 0 to 1.

For example, the air can be 20% cool and, at the same time, not cool. This means that in the case of the traditional decision tree, the boundaries used are exact, but in the case of fuzzy decision, the boundaries used are curved [8].

## 2.3 Proposed Approach for Document Clustering

The proposed fuzzy algorithm is used as a text document clustering tool in this study. In this section, the general architecture for the process of text document clustering is synthesized as well as the fuzzy proposed algorithm and its learning.

The classification of text documents contains several steps. The process is presented as follows (Figure 2.1).

### 2.3.1 Collecting Documents

First, we collect a set of documents in different domains, such as biological documents, mathematical documents, and others. To do so, we propose to use the JabRef software. It is a program that provides an interface for importing data and for managing files.

JabRef lets us build our own biological pertinent abstracts database (Figure 2.2).

### 2.3.2 Processing Documents

The selection step consists of the extraction of pertinent abstracts from the MEDLINE database. At this step, we must guarantee that all abstracts are randomly selected and without any user intervention.

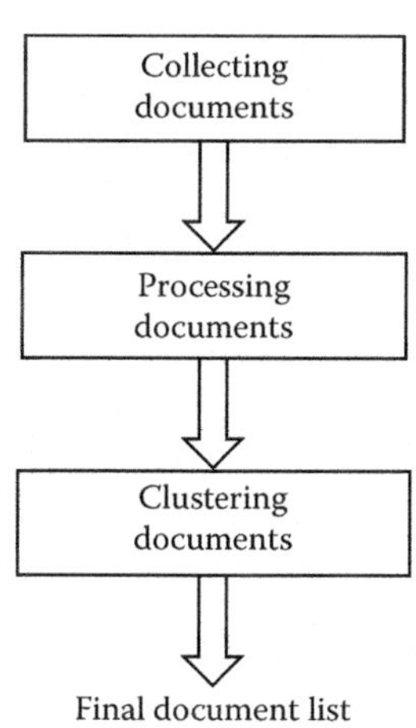


**FIGURE 2.1**
General architecture for the process of text document clustering.

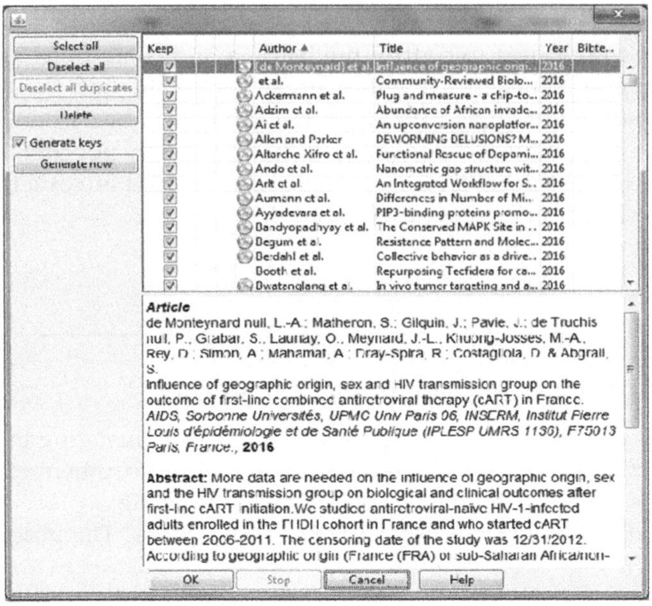

**FIGURE 2.2**
Collecting documents.

Our selected documents are defined by two tasks: the cleaning and the vector representation task.

### *2.3.2.1 Cleaning Documents*

The cleaning step is characterized by the removal of useless information, also known as stop words. This step consists of decomposing a document into separate words and deleting the unnecessary ones. It is used to reduce the textual data size and improve efficiency and effectiveness. An example of English stop words is shown in Figure 2.3.

We give an example that describes the cleaning step (Table 2.1) in more detail.

The first column represents the abstract without deleting of unnecessary words, but the second represents the abstract after the cleaning step.

### *2.3.2.2 Vector Representation of Documents*

The representation step is characterized by modeling the document as a vector. In our work, we decide to use vector representation thanks to its ability to facilitate the calculation of terms. These vectors contain the document terms associated with their frequency.

An example of the vector representation step is given in Table 2.2.

The first column represents the abstract after the cleaning step, but the second column represents the vector representation process.

For example, the word "biological" appears just once in the abstract and the word "transmission" appears twice in the abstract.

| | | | |
|---|---|---|---|
| a | couldn't | ourselves | they've |
| about | did | out | this |
| above | didn't | over | those |
| after | do | own | through |
| again | doesn't | she | to |
| against | doing | she'd | too |
| all | don't | she'll | under |
| am | down | she's | until |
| an | during | should | up |
| and | each | so | very |
| any | few | some | was |
| are | for | such | wasn't |
| aren't | from | than | we |
| as | further | that | we'd |
| at | had | that's | we'll |
| be | have | the | we're |
| because | haven't | their | we've |
| been | having | theirs | were |
| before | he | them | weren't |
| below | he'll | then | what's |
| between | he's | there | when |

**FIGURE 2.3**
Example of English stop words.

**TABLE 2.1**
Example of Text Cleaning (Biological MEDLINE)

| Input Text | Output Text |
|---|---|
| More data are needed on the influence of geographic origin, sex and the HIV transmission group on biological and clinical outcomes after first-line cART initiation. We studied antiretroviral-naïve HIV-1-infected adults enrolled in the FHDH cohort in France and who started cART between 2006–2011. | Data needed influence geographic origin, sex HIV transmission group biological clinical outcomes first-line cART initiation studied antiretroviral naïve HIV-infected adults enrolled FHDH cohort France started cART. |

**TABLE 2.2**
Example of the Vector Representation Step

| Input Text | Output Text |
|---|---|
| Data needed influence geographic origin, sex HIV transmission group biological clinical outcomes first-line cART initiation studied antiretroviral naïve HIV infected adults enrolled FHDH cohort France started cART. | (Data,1), (needed,1), (influence,1), (geographic,1), (origin,1), (sex,3), (HIV,2), (transmission,1), (group,1), (biological,1), (clinical,1), (antiretroviral,1), (naive,1), (infected,1), (adults,1), (enrolled,1), (FHDH,1), (cohort,1), (France,1), (started,1), (cART,3) |

In our approach, we decided to use the representation vector to transform a document from a set of terms to a vector that contains each term associated with its weight. The vector is represented as follows:

$$V_j = \begin{pmatrix} \coprod_{dj}(t1)\coprod_{dj}(t1), \coprod_{dj}(t2)\coprod_{dj}(t2) \\ \coprod_{dj}(t3)\coprod_{dj}(t3), \ldots, \coprod_{dj}(ti)\coprod_{dj}(ti) \end{pmatrix},$$

where

$V_j$ is the vector that represents the document j
$d_j$ is the document j
$t_i$ is the term i
$\coprod_{dj}(ti)\coprod_{dj}(ti)$ is the membership degree of the term i in the document j

The membership degree is defined by measuring the frequency of the term t in the document. The representation process consists of the transformation of each document to a vector. A set of documents can be represented by a matrix (Figure 2.4).

The rows represent the words that appear at least once, and the columns represent the clusters.

### 2.3.3 Clustering Documents

In our work, we decide to cluster the text document with a fuzzy clustering algorithm. Clustering with the fuzzy clustering algorithm is unsupervised and provides a cluster without a priori known number of classes, which means that we have to choose arbitrary k clusters.

Documents will be compared by the number of terms that appear in the text, and each document will be associated with the nearest neighbor cluster.

The fuzzy clustering algorithm is given as follows:

i. Choose arbitrary k clusters.
ii. Calculate cluster centers.
iii. Calculate membership degrees of terms in documents.
iv. Repeat steps (ii) and (iii) until emergence is stable.

Number of clusters

| | $T_1$ | $T_2$ | T... | $T_n$ |
|---|---|---|---|---|
| $C_1$ | | | | |
| $C_2$ | | $\coprod_{Dz}(t2)$ | | |
| C... | | | | |
| $C_n$ | | | | |

**FIGURE 2.4**
Membership matrix.

For more details:

i. k can take values from 3 to 5,7,…, but it should be an odd number.
ii. After finishing the vector representation process, we calculate the center's vectors to facilitate the classification step. The center vector is defined as follows:

$$C_j = \frac{\sum_{i=1}^{N} \coprod_{Cj}(ti) * ti}{\sum_{i=1}^{N} \coprod_{Cj}(ti)}$$

where
$C_j$ is the center
$\coprod_{Cj}(ti)\coprod_{Cj}(ti)$ is the membership degree of the term i in the cluster j

iii. A similarity measure between two documents is computed. In this study, the Euclidean distance between two documents di + 1 and di is computed as follows:

$$D(X,Y) = \left[(X_1 - Y_1)^2 + (X_2 - Y_2)^2 + \cdots + (X_n - Y_n)^{1/2}\right]$$

We used the Euclidean distance to calculate the similarities and dissimilarities between vectors. In our work, this metric represents the distance separating the document i from the document j.
iv. If $\| U^{(k+1)} - U^{(k)} \| < \varepsilon$ then stop; otherwise, repeat steps (ii) and (iii) until emergence is stable.

## 2.4 Experimentation and Test

To evaluate the effectiveness of the proposed fuzzy algorithm, as a text document clustering tool, was conducted. In our work, the abstracts of 100 randomly selected articles from JabRef were used for experimentation to evaluate the classification process.

A predefined list of stop words, composed of 238 words, was used. The parameters used were k = 3 clusters. In this study, a laptop with an Intel Core i2 processor and Windows (64-bit) was used. All our experiments were implemented using the Java language and compiled in the Eclipse framework.

To validate our approach, we need to evaluate its implementation and we also need to test it on a set of adequate data.

First of all, we extracted all the documents related to the biological MEDLINE. For this reason, we used JabRef as shown in Figure 2.5, where JabRef imported abstracts from the MEDLINE database.

To calculate the performance of our proposed approach, we decided to use the classification rate.

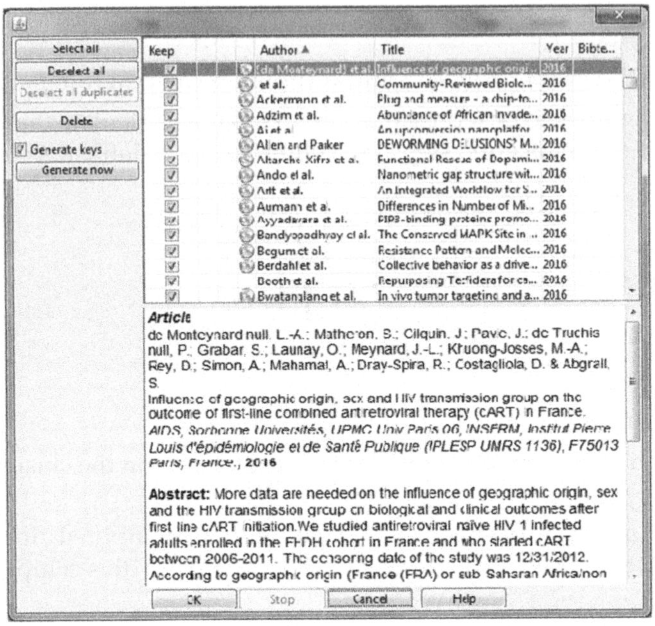


**FIGURE 2.5**
Abstract selection from MEDLINE.

The implementation of classification should be evaluated according to standard measurements; to do so, we chose the classification rate. It measures the system efficacy by calculating the ratio of documents correctly classified to the total number of samples in the test set.

$$\text{Classification rate} = \frac{\text{The number of documents correctly classified}}{\text{The total number of samples in the set}}$$

For experimentation, we used the abstract extracted from the JabRef system. So for the biological documents extracted, we measured the accuracy rate using both our fuzzy classification algorithm and the C4.5 algorithm (Table 2.3).

Table 2.3 presents our test results for the biological MEDLINE.

The results are encouraging since the classification rates obtained are comparable.

On one hand, the C4.5 algorithm showed very good results thanks to its ability to classify all the documents, but it was unable to detect the similarities between them.

On the other hand, our proposed algorithm showed its efficiency in classifying all the documents, but it also has an important characteristic over C4.5, which is its ability to detect similar documents.

**TABLE 2.3**
Classification Rates of the Biological MEDLINE

| Data Set | Algorithm | Accuracy (%) |
|---|---|---|
| Biological documents | C4.5 | 70 |
| | Fuzzy classification algorithm | 72.1 |

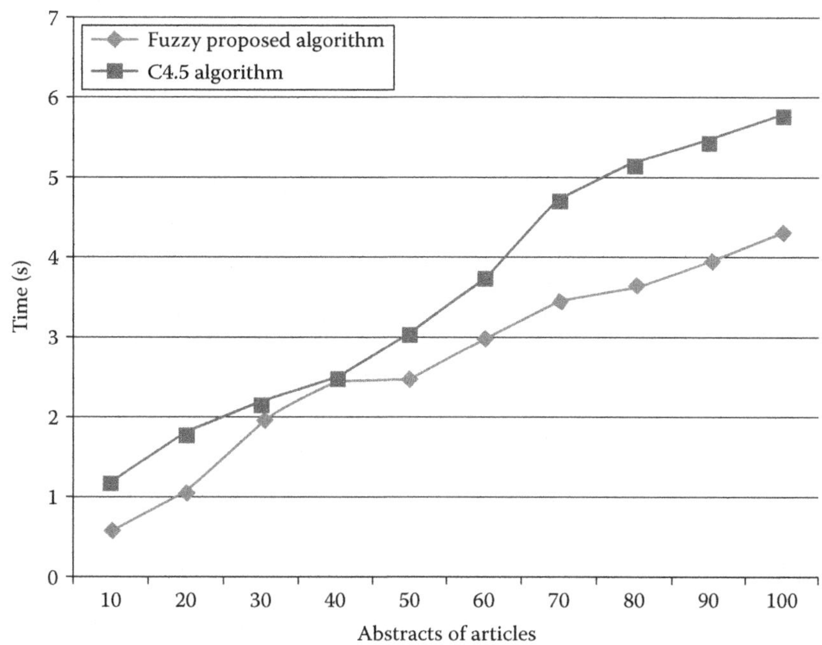


**FIGURE 2.6**
Runtime curve for biological MEDLINE.

In Figure 2.6, we used the Biological MEDLINE. To draw the dark gray curve, we began by testing, or evaluating, our algorithm for only 10 documents, and we recorded the corresponding run time. Then, we incremented our tested documents by 10 each time until a total of 100 documents were completed. We followed the same steps to draw the light gray curve using C4.5.

We noticed, according to Figure 2.6, that the proposed algorithm takes lesser time for running than C4.5 regardless of the size of the data set.

After measuring the classification rates and computing the running times for the two different algorithms, C4.5 and the fuzzy proposed algorithm, we will use the fuzzy proposed algorithm with a different number of clusters to show the tree size and the classification rates.

For example:

Number of clusters = 3, tree size = 5, classification rate = 95%

Number of clusters = 5, tree size = 7, classification rate = 93%

Number of clusters = 15, tree size = 27, classification rate = 73%

The results show that the classification rates decreased as the number of clusters increased and the tree size increased as the number of clusters increased.

Figures 2.7 and 2.8 summarize the time required for the classification process, where a new abstract was provided for the number of clusters 3, 5, and 15, respectively.

With the proposed approach, articles imported from the JabRef software could be clustered and visualized as a list of text documents. It was expected that the computational complexity would increase with increasing the number of documents used for clustering.

Nevertheless, the proposed approach constitutes a useful tool for users in the clustering of text documents.

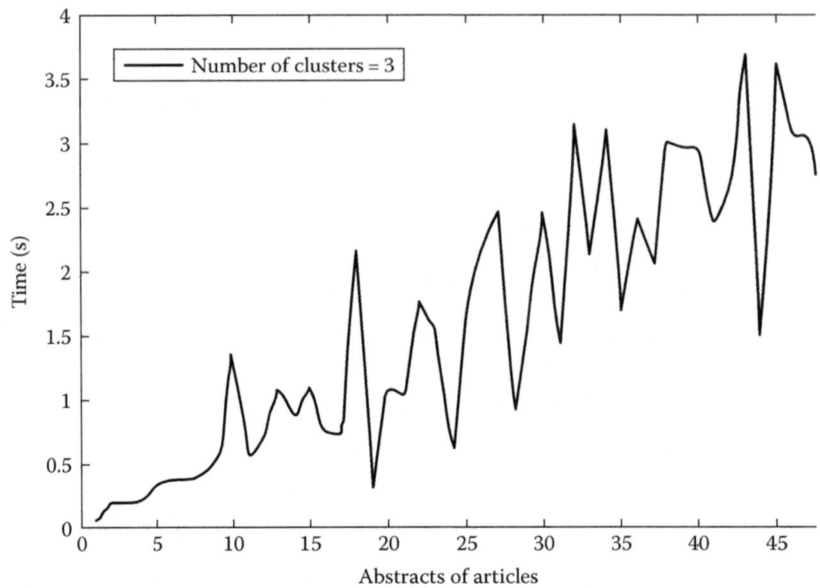


**FIGURE 2.7**
Learning time with number of clusters = 3.

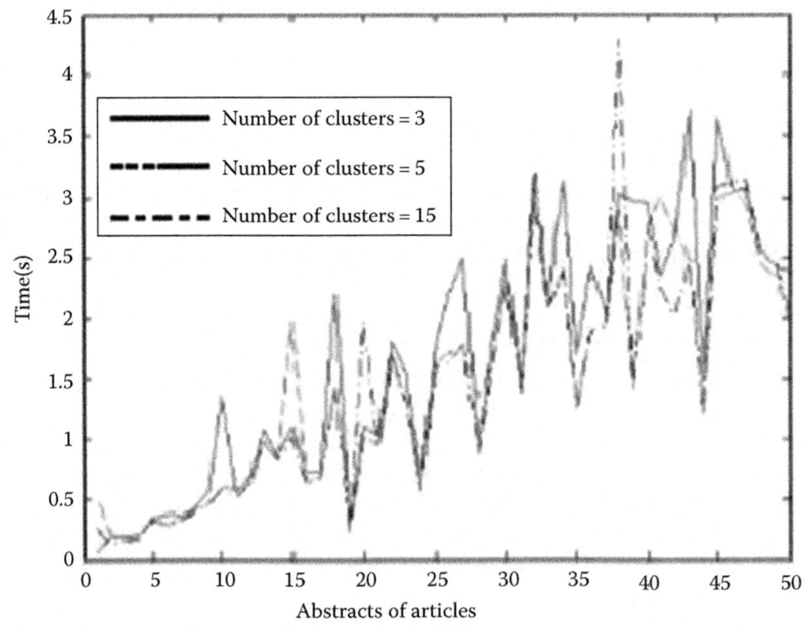


**FIGURE 2.8**
Learning time with number of clusters = 3, 5, 15.

The goal of this section is to evaluate the proposed approach and to validate it for classifying text documents. We used the JabRef system for the extraction of a text document and the Eclipse for the implementation, and we reported the results.

In an experimental study, we used the data set grouping documents related to the biological MEDLINE. Then, we compared the fuzzy classification algorithm with the C4.5 algorithm by measuring the classification rates and the running times.

The results we obtained are promising. They showed that the fuzzy classification algorithm is promising. They prove its efficacy not only to classify documents but also to detect their similarities. Moreover, the obtained results show that the complexity in terms of time increased with increasing the number of articles used for clustering and also with increasing the number of clusters.

## 2.5 Conclusion

Nowadays, getting the right information at the right time has become a necessity. However, upon the classification of a text document, a classifier may not be able to determine the class associated with the document. Therefore, our need for an application to organize and to classify text documents is becoming more and more crucial.

In this work, we proposed a new approach for detecting similarities and classifying documents. It is based on fuzzy logic.

Fuzzy logic has become a very important field thanks to its ability to search for the similarities between documents.

In this approach, we used the vector space model to represent documents, and we also used the known Euclidean distance to measure the similarities between them. Then, we elaborated a new method that combines fuzzy logic and the distance metric in order to detect similar documents and classify them.

We used the data set grouping documents related to the biological MEDLINE. Then, we compared the fuzzy classification algorithm with the C4.5 algorithm by measuring the classification rates. The obtained results are promising. They showed that the fuzzy classification algorithm is competitive. They prove its efficacy not only to classify documents but also to detect their similarities.

However, the measurement of the learning time depends on the choice of cluster numbers and distance metric. In future works, we aim to improve our proposed fuzzy algorithm by the enhancement of computing the cluster centers, by the choice of cluster numbers used for the classification process, and by the choice of the distance metric.

## References

1. Bart, K. and Satoru, I. (1993). Fuzzy logic, retrieved from http://Fortunecity.com/emachines/e11/86/fuzzylog.html. Scientific American, Vol. 269, July 1993.
2. Quinlan, J. (1993). *C4.5: Programs for Machine Learning*. Morgan Kaufmann Publishers, San Francisco, CA.
3. Chang, W. L., Tay, K. M., and Lim, C. P. (2014). An evolving tree for text document clustering and visualization. In *Soft Computing in Industrial Applications* (pp. 141–151). Springer International Publishing.

4. Kohonen, T. (2001). *Self-Organizing Maps*, vol. 30. Springer series in information sciences. Springer, Berlin, Germany.
5. Zadeh, L. A. (1965). Fuzzy sets. *Information and Control*, 8(3), 338–353.
6. Zimmermann, H. J. (1996). Fuzzy control. In *Fuzzy Set Theory—and Its Applications*. Springer, Dordrecht, the Netherlands, 59, 203–240.
7. Hüllermeier, E. (2005). Fuzzy methods in machine learning and data mining: Status and prospects. *Fuzzy Sets and Systems*, 156(3), 387–406.
8. Dash, S. K., Mohanty, G., and Mohanty, A. (2012). Intelligent air conditioning system using fuzzy logic. *International Journal of Scientific and Engineering Research*, 3(12), 1–6.

# 3

# *Toward Modeling Semiautomatic Data Warehouses: Guided by Social Interactions*

**Wafa Tebourski, Wahiba Ben Abdessalem Karaa, and Henda Ben Ghezela**

**CONTENTS**



**ABSTRACT** Due to the growing interest in social network activity, massive volumes of user-related data and a new area of data analysis have emerged. The data warehouses are used to support decision-making activities in a system of business intelligence to guarantee the structure and analysis of multidimensional data. The data are presented in the form of simple data marts (star schema), involving the well-identified collection of facts and dimensions.

In this chapter, we propose a novel approach to design data warehouses from data marts using a descriptive statistical technique for the analysis of multidimensional data in the principal components analysis (PCA) in the medical social network.

**KEY WORDS:** *medical social network, design of data warehouse, data marts, mixed approaches, principal components analysis (PCA), statistical analysis, correlation.*

## 3.1 Introduction

A data warehouse is "a collection of data, integrated, nonvolatile and historized for decision making"[1]. Recently, data warehouses have become critical components of business intelligence. They have been effectively implemented in various sectors. Indeed, data warehouse merge and standardize databases, allowing analysis and decision making.

In fact, the exploration for a method of data warehouses modeling has become track successfully. In the literature, several approaches for data warehouses design are proposed. These approaches can be classified into three categories: (i) approaches directed by the sources (bottom-up), (ii) approaches directed by the needs (top-down), and (iii) mixed approaches. It should be noted that the bottom-up and top-down approaches suffer from some limitations; in contrast, mixed approaches are based on requirements and sources with the same importance has better feasible consequences.

Based on mixed approaches, and in order to obtain clear advantages, this chapter studies a new approach to data warehouse design, based on the principal component analysis (PCA) in the medical social network, called DWDAPMSN (Data Warehouse Designing Approach based on Principal component analysis in Medical Social Network). The advantage made by this approach is its focus on the statistical basis because the PCA is a descriptive method that intends to summarize the variables in a multidimensional decreased number of. In fact, this analysis offers many advantages such as performance, flexibility, and mathematical simplicity during the implementation phase. The rest of the chapter is organized as follows. In Section 3.2, we present a state of the art of the different approaches that have been used to design data warehouses. In Section 3.3, we introduce our approach and present a case study on medical data to illustrate the proposed model. Finally, we conclude and give some future work in Section 3.4.

## 3.2 State of the Art

### 3.2.1 Approaches to Designing Data Warehouses

A great number of researchers have concentrated on modeling data warehouse diagrams. Our objective in this section is to compare the various methods using many measures.

#### *3.2.1.1 Sources-Based Approaches*

Modeling information storing depends on deep study of the information model, normally, the entity relationship model (E/R). These methods facilitate the way of ETL (extraction transformation-load) every entity and relation in the source model will be represented by multidimensional concepts.

We can cite as examples different works in this domain [2–7]. The bottom-up approaches have many difficulties like the weakness in making a decision by the decision maker. The outcome may be schemas that do not satisfy users' needs.

*3.2.1.1.1 Golfarelli et al.'s Approach*

The authors [3] propose a formal model: dimensional fact model (DFM), which is a multidimensional graphical model, differencing concepts such as facts, measures, dimensions, and hierarchies. This model is in the form of a tree made of examples or patterns.

In this method, extracting a pattern has two stages:

1. Stating facts
2. Building element in the form of a tree: to exclude the wrong elements and to study the needed measures and the classes of each fact

*3.2.1.1.2 Hüsemann et al.'s Approach*

This work uses a method that has four stages [2] like analysis and specification needs.

The researchers select the needs element for a model E/A source and identify the use of a measure of fact or dimension. Extra-needs are added by complex measures:

1. The conceptual modeling: at this stage, a conversion of the semiformal arrangement requirements of the multidimensional conceptual schema is accomplished.
2. The logical modeling: this stage transforms the schema as an idea into the real design.
3. The physical modeling: this stage concretizes the use of diagrams.

*3.2.1.1.3 Romero et al. Approach*

The aim of this method [4] is to organize the multidimensional concepts from domain ontology. This approach constitutes four principles of how to use ideas with many measures: (i) the multidimensional model, (ii) the constraint of multidimensional space arrangement, (iii) the integrity of constraint base, and (iv) of the additive constraint.

#### *3.2.1.2 Requirements-Based Approaches*

The top-down approaches are those that show the stages of the designing requirement specification in the form of conceptual schemas. Such approaches attempt to limit the decisional information system failure risk. Therefore, a lot of researches have been conducted in order to create requirements-based data warehouses such as [8–13].

In the requirements-based approaches, more generated models may not be satisfactory because the available data sources are complex and heterogeneous, which makes the ETL process more difficult to achieve.

*3.2.1.2.1 Kimball's Approach*

This approach [13] is a requirement-based approach designed to obtain a logical design of a data warehouse. This method is informal, using a full multidimensional concept leading to a multidimensional schemas guide. The proposed method has two steps:

1. *The bus designing*: This step aims to recognize all the data marts that the designer pursues to build. The dimensions of each data mart should be classified. Thus, an ad hoc matrix is constructed to restrict the multidimensional needs and point out associations between different data marts.
2. *The cycle of multidimensional*: This step is defined by five stages: (i) making a plan, (ii) explanation of business needs, (iii) selection of technology, (iv) designing information, and (v) practical use.

*3.2.1.2.2 The Approach of Cabbibo and Torlone*

This approach generates a logic diagram of ER (entity relationship or n-ary) diagrams [9]. Furthermore, it can generate multidimensional schemas in the form of relational databases or multidimensional matrices. However, this approach did not have clear rules. Nonetheless, this approach has introduced the basic foundations applied, later, to the rest of the methodologies.

The proposed method consists of four stages. The first and the second stages permit reorganization of facts and dimensions identification, as well as the ER diagram. The third and the fourth steps provide the multidimensional diagram.

*3.2.1.2.3 Mazôn et al.'s Approach*

This work presents a requirement-based approach. The aim of this proposition is to merge the business objectives [8] in the specification of the requirements using i* technology.

This approach depends on needs and it has three steps. (i) Explaining business aim, which involves specifying the main objectives of a company. These goals can be classified into three abstraction levels: strategic, informational, and decision-making. (ii) Designing needs using i* technology; a step that identifies the users of data warehouse, the business goals of the organization, and the relationship between these two elements. (iii) Converting the different measures into objectives [9].

*3.2.1.2.4 Giorgini et al.'s Approach*

This approach [10] begins with the specification of requirements and carries out two models: organizational model and decision-making model. After that, a construction step is accomplished. The objective is to present a conceptual model that results from the relational model in a decision-making perspective and, afterward, is refined using the hierarchies of the organizational model diagram.

### 3.2.1.3 Mixed Approaches

The mixed approaches combine needs and sources. They include both bottom-up and top-down approaches in order to benefit from their advantages [11,14,15].

*3.2.1.3.1 Bonifati et al.'s Approach*

The varied methods are composed of requirements and sources. Bonifati et al. studied the semiautomatic approach [14], which is made of the two elements: requirements and sources.

This method has three stages: (i) Bottom-up analysis: to examine the E/R model of the data source and construct star schemas candidates based on some patterns. The bottom-up analysis can create a great number of candidate patterns. The authors proposed an algorithm that transforms each *n–m* association into *1–n*, by considering the E/R model as a graph. (ii) Top-down analysis: to collect the analyzed requirements and refine and aggregate them in a tabular report of abstraction. This step has as outputs, the star schemas principles. (iii) Integration: to match each pattern from the top-down analysis, with all the candidates, created by the bottom-up analysis, in order to meet the requirements of decision diagrams.

*3.2.1.3.2 Nabli et al.'s Approach*

This approach [15] initiated a mixed method of using the automated design of data marts and data warehouse, beginning with semistructured OLAP needs expressed in a tabular form. The approach encompasses three steps: (i) the acquisition of OLAP requirements, performed using ontologies; (ii) the generation of data marts patterns; and (iii) the generation of the data warehouse schema.

*3.2.1.3.3 Giorgini et al.'s Approach*

The authors introduced a mixed approach based on three stages: (i) requirements analysis—a step that generates a decision model and an organizational model; (ii) matching needs with sources—a step in which the decision model is mapped to an E/R data source, through the organizational design; and (iii) refinement [11]—wherein the multidimensional design is enhanced through the structure of hierarchies and their refinement.

#### 3.2.1.4 Comparative Study

Table 3.1 illustrates the comparison between the diverse approaches of data warehouses design based on different criteria. The majority of these methods permit creating data warehouse as the work [2,3,8–11,14]. Nevertheless, the rest leads to generating data marts of such work [13,15]. All these approaches are formal methods except the approach of [9,13], where an informal modeling is introduced. The design of a data warehouse consists of several data sources, namely, the relational schema such as those in the works of [2,3,8,10,11,13,14]. Other methods [4,15] use ontology as data sources. Given the complexity of the requirement specifications, some researchers have developed, using different techniques and methods such as i* technique [8], the TROPOS method [10,11] and the GQM method [14].

Based on this comparative study, we can conclude that sources-based approaches are functional if the diagram of the data source is simple and available. In this category, they suffer generally from needs engineering patterns. However, the requirements-based approaches, focus on the needs specification, which are, habitually changeable and restrictedly expressed. Consequently, the data warehouses model cannot be totally based on data sources or requirements. Certainly, both ascending and descending approaches are complementary, since they can be mixed together to give better results, being the subject of the third approach called mixed approaches.

### 3.2.2 Social Network

Social networking sites give tools for people to connect and collaborate online. They are growing rapidly in popularity. Social networks allow their users to interact and provide data. This results in huge data collections that provide information on the attitudes of people. Indeed, the analysis of data from social networks has become a huge interest in research. In recent years, the explosion of social networking activity has given rise to a new range of data analysis based on user data from email, blog, forum and articles, and so on. Today, the company is looking to boost their customers and improve their services and marketing with public relations that are embedded in social networks such as advertising, suggestions, recommendations, and so on.

**TABLE 3.1**

Comparison between Diverse Approaches to Data Warehouse Design

| Approaches | | Approaches Directed by Sources | | | Approaches Directed by Requirements | | | | Mixed Approaches | | |
|---|---|---|---|---|---|---|---|---|---|---|---|
| **Criteria** | | **Golfarelli et al.** | **Hiserman et al.** | **Romero et al.** | **Kimball et al.** | **Cabibbo et al.** | **Mazon et al.** | **Giorgini et al.** | **Bonifati et al.** | **Nabli et al.** | **Giorgini et al.** |
| Goals | Data warehouse | + | + | + | – | + | + | + | + | – | + |
| | Data marts | – | – | – | + | – | – | – | – | + | – |
| Modeling | Formal | + | + | + | . | . | + | + | + | + | + |
| | Informal | – | – | – | + | + | – | – | – | – | – |
| Type of data sources | ER diagram | – | – | – | – | + | – | – | – | – | – |
| | Ontology | – | – | + | – | – | – | – | – | + | – |
| | Relational schema | + | + | – | + | – | + | + | + | – | + |
| Methods used for requirements specification | | – | – | – | – | – | Technique i* | TROPOS | GQM (goals/question) | – | TROPOS |

*Source:* Tebourski, W. et al., *Int. J. Comp. Sci.*, 10(5), 2, 2013.

## 3.3 New Approach for Data Warehouse Design Based on Principal Component Analysis in Medical Social Network

In this section, our functional architecture is presented.

### 3.3.1 Functional Architecture

Figure 3.1 summarizes the three steps of our approach. The first step is the specification of OLAP requirement based on TALN tool. For the upcoming steps, the user declares its analytical objectives (facts). Beginning with a group of starters like social networks, the first step is to perform preprocessing on qualitative data stored in tables. The approach examines the similarity between the data, based on the correlation between variables, in order to group data into factors. These factors are translated into two types: (i) measures' factors, which include heterogeneous data and (ii) dimensions' factors, which support homogeneous data. Thereafter, the proposed model studies the correlation between the obtained facts and factors. The final stage is the diagram generation of data marts that will be validated by the expert. The third step is to generate a data warehouse schema from data marts schema with the star join schema.

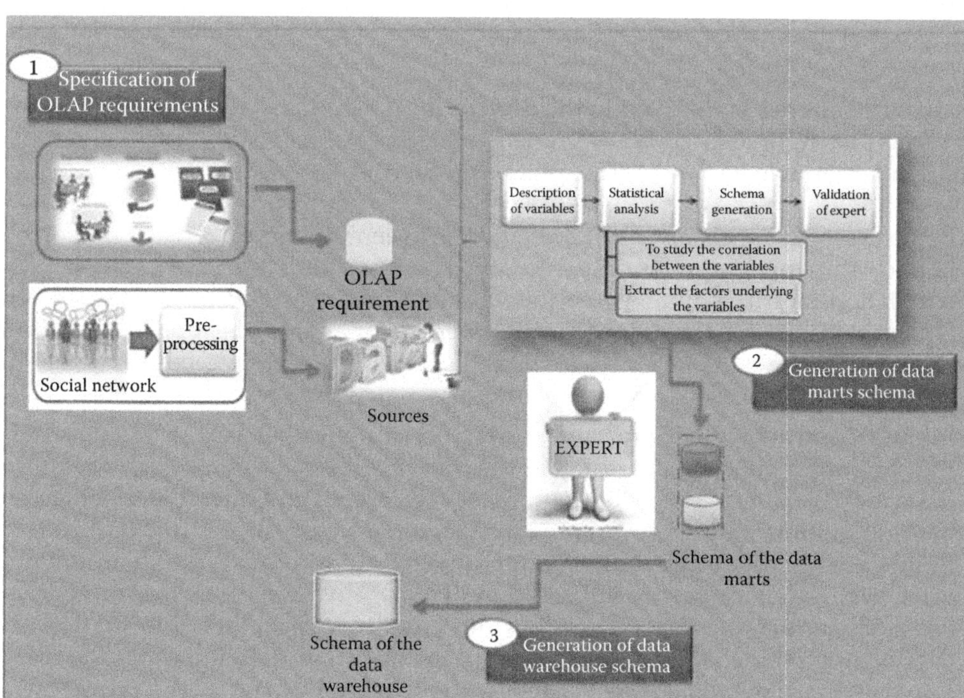


**FIGURE 3.1**
Functional architecture.

### 3.3.2 Process

The process driving our DWDAPMSN approach has three steps. The first step presents the specification of OLAP requirement. The second step has a further four steps, which will be presented in the following sections, and the third step presents the generation of data warehouse schema.

#### *3.3.2.1 Step 1: Specification of OLAP Requirement*

This step consists in specifying requirements based on automatic processing in NLP (natural language processing). Generally, pretreatment of text follows a sequence of steps: (i) sentence splitting, (ii) tokenization, (iii) post tagging, and (iv) syntactic parsing. Each step is designed to solve problems at this level of processing power and the upper level with a flow of accurate data [17].

#### *3.3.2.2 Step 2: Generation of Data Marts Schema*

##### *3.3.2.2.1 Description of Variables*

To describe the studied variables, we propose to convert qualitative data into quantitative ones, based on one of two matrices: (i) Contingency matrix: to cross two unimodal variables; if the co-occurrence measure is applied to both unimodal variables, we speak of a contingency measure. (ii) Co-occurrence matrix: takes several representations, regarding the purpose of the analysis. It is used to perform quantitative relational analysis. (iii) Presence–absence matrix: to make the junction of two variables. This requires the existence of at least one individual.

##### *3.3.2.2.2 Statistical Analysis*

Our statistical analysis is reflected in the drop of the data organized into a set of factors by minimizing the residual variance (intraitems variance) and maximizing the cumulative variance (intervariable variance). This phase consists of two stages:

1. In the first stage, we search the correlation between the variables using the correlation or the covariance matrices to study for synthetic variables. In fact, the correlation matrix is used when the variables are measured on different scales, whereas we use the covariance matrix when factor analysis is applied to multiple groups with different variances for variables. The data matrix needs to contain the appropriate correlations, to validate the association between the variables.
2. In the second stage, we determine the factors' numbers and extract some criteria that are often chosen according to the variance of each item. Each factor can describe (i) Rule Kaiser–Gutman, which is an eigenvalue that denotes the amount of information represented by a factor. (ii) Test elbow: This test is essentially based on the factors' eigenvalues, in a relative context (not in an absolute one). When the amount of information between two factors is low or zero, we can deduce that the last factor does not contain satisfactory information and cannot be a retained variable. (iii) Percentage of variance: It illustrates the cumulative variance percentage, extracted by the successive factors. The purpose is to guarantee that a factor explains a significant amount of variance. It is suitable to stop the extraction of factors when 60% of the explained variance is already extracted [18].

*3.3.2.2.3 Schema Generation*

All components that are extracted from the candidate data marts schema and which contain the fact summarizing the subject of analysis will include the dimensions that form the axes of the topic analysis.

*3.3.2.2.4 Validation of Expert*

Our proposal is iterative and incremental. In fact, the expert can validate the generated schema or confirm and restart another iteration of the process to accomplish more satisfactory results.

#### ***3.3.2.3 Step 3: Generation of Data Warehouse Schema***

In this step, (i) we identify common dimensions between the star schemas; (ii) we present a data warehouse schema with the star join schema; and (iii) finally, the expert can validate the generated constellation schema.

### 3.3.3 Algorithm

Our new approach called DWDAPMSN takes as input all the data set after data processing task. The outputs are the factors, recapitulating the most correlated variables, from which data mart schemas will be extracted. Consequently, the data warehouse schema is generated from the data marts. The used notations of DWDAPMSN algorithm are presented in Table 3.2 and the pseudo-code of this algorithm is illustrated in the subsequent section. DWDAPMSN contains two essential functions:

The first function recapitulates the data for the most explanatory variables. Then, it calculates the correlation between these variables; the result is a correlation matrix. In a second stage, the factors are extracted, based on the correlated variables. At this stage, we compute the total variance, reflecting the degree of information and including all the variables.

The second function uses the first function to compare the generated components and extract those shared. Finally, the expert can intervene to identify the obtained components and to validate the data marts schema. In the third stage, the data warehouse schema is generated and then validated by an expert (Figure 3.2).

**TABLE 3.2**

List of Used Notations

| Notation | Description |
|---|---|
| Cj | Component j |
| C | All components |
| $C_{SI}$ | Shared component item |
| Corr (k,l) | Correlation between the two variables k and l |
| DS | Data Source |
| LstC | List of components |
| Mcorr | Correlation matrix |
| nC | Number of components |
| Var (m,o) | Cumulative variance of m and o |
| Vt | Total variance |

```
Function Generate C (Data d)
Begin:
//Study the correlation between variables
    For(k = 1; k < n; k++)do
        For(l = 1; l < n−1; l++)do
            Mcorr = Calculate the correlation matrix
            Corr(k,l) = calcul_correlation (k,l)
            Store Corr(k,l) in Mcorr
//Retrieve the factors underlying variables
    For(m = 1; m < n; m++)do
        Vt = Calculate the cumulative variance
        Var(m,o) = calcul_variance
        Store Var(m,o) in Vt
//Identify C from Vt
    For(i = 1; i < n; i++)do
        For(j = 1; j < n; j++)do
            If Vt(i,Cj) is maximum then
                Affect i in Cj
```

```
Return C
//Return list of components including shared components
Function Generate LstC
    For each data d in DS
        C = Generation C (d)
        LstC.add(C)
For(int i = 1; i < length(Lst);i++)do
    For(int j = 2; i < length(Lst);j++)do
        C = LstC(i)
        C' = LstC(j)
        For (int k = 1; k < length(LstC);k++)do
            For (int l = 1; l < length(LstC); l++)do
                If(C(k) ≡ C'(l))
                    Affect C'(l) in CSI
Return LstC, CIS
End
```

**FIGURE 3.2**
Algorithm DWDAPMSN: *D*ata *W*arehouse *D*esigning *A*pproach based on *P*rincipal Component *A*nalysis in *M*edical *S*ocial *N*etwork.

### 3.3.4 Case Study

The case study that concerns our designed data warehouse aims to analyze the different subjects presented by doctors and assess their importance through the number of discussions around a particular topic. A part of the involved medical social network data warehouse is shown in Figure 3.3.

Each medical specialty includes areas and subspecialties of very specific skills.

#### *3.3.4.1 Step 1: Specification OLAP Requirement*

In this step, we have employed the Gate tool to specify OLAP requirement, as shown in Figure 3.4

#### *3.3.4.2 Step 2: Generation of Data Marts Schema*

Several types of social media exist. In this chapter, we use an example of a medical social network that we built in order to evaluate our contribution.

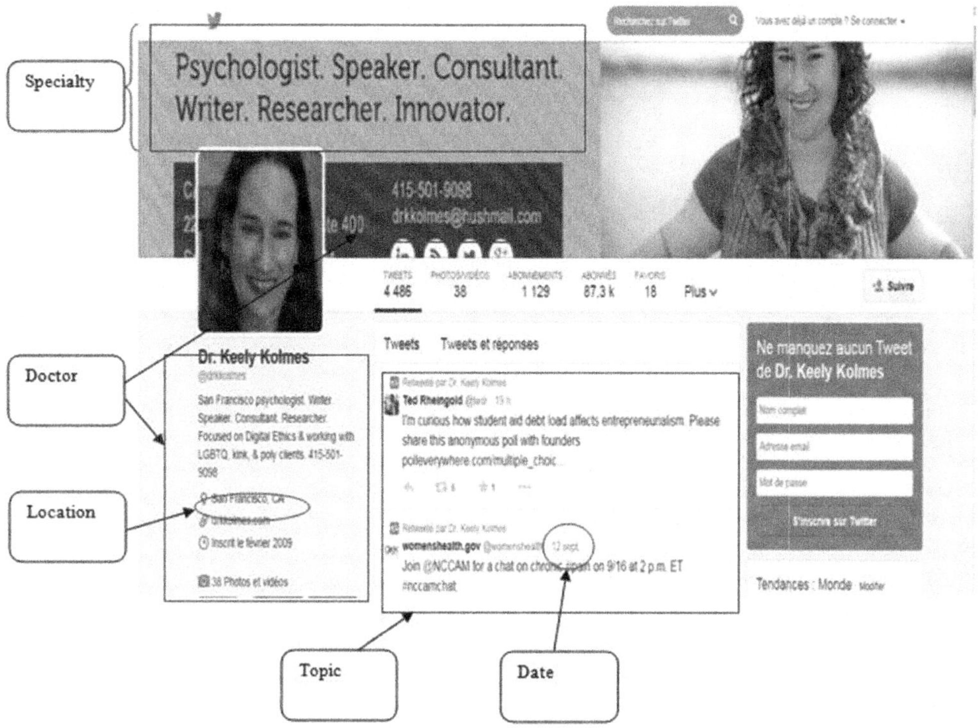


**FIGURE 3.3**
An illustration of our process for building medical social network warehousing.

*3.3.4.2.1 Generation of "Topic" Data Marts Schema*

- *Description of variables*: Twitter is a platform adapted to the new needs of communication and collaboration in an independent environment. It enhances the skills of doctors and allows them to communicate, recommend, and access the latest global medical innovations. The variables of our dataset are taken from a medical social network via the platform Twitter:
  - Id : doctor code.
  - Tel : the phone of doctor.
  - Specialty: the specialty of doctor, 1 for general practice, 2 for anesthesia–reanimation, 3 for general medicine, 4 for carcinology surgery, 5 for cardiology, 6 for general surgery, 7 for orthopedic surgery, 8 for gastrointestinal pathology, 9 for infectious diseases, 10 for hematology, 11 for gynecology-obstetrics and 12 for endocrinology.
  - Location: the country of the doctor.
  - Date: date of the discussion.
  - Topic: text of discussion.
- *Statistical analysis*: The statistical analysis is presented in the following sections:
  - Study of the correlation between variables: The aim of this step is to reduce the data by forming a smaller number of more variables and more correlations. For this end, we use the correlation matrix that contains all the correlations between

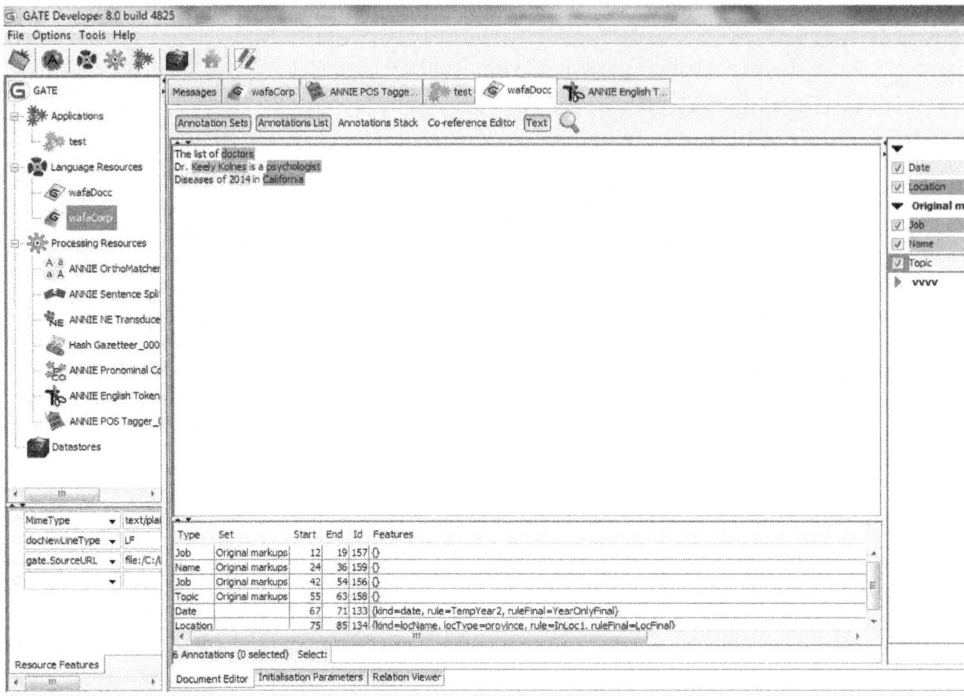


**FIGURE 3.4**
Requirement annotation using Gate API.

variables (Table 3.3). The "specialty" variable is positively correlated with the variables "Id." "Location" and "Date" variables are correlated with the "Topic." variable. This correlation between variables will define the set of variables that will constitute the set of factors (components). All the correlated variables will be assembled into separate factors.

- *Extract factors using variables*: It is the total explained variance. The total variance defines the information level represented by each component or each factor. Table 3.4 shows that there are five input variables used in order to identify four components. The first component encloses 30.32% of the total information of all variables. The second comprises 51.76%, the third 72.55%, and the fourth contains 92.30%. Generally, we choose the variables that have total information

**TABLE 3.3**
Correlation Matrix of Our Dataset

| | | Id | Specialty | Location | Number_discussion | Date |
|---|---|---|---|---|---|---|
| Correlation | Id | 1.000 | 0.554 | 0.232 | −0.253 | 0.069 |
| | Specialty | 0.554 | 1.000 | −0.101 | −0.261 | 0.577 |
| | Location | 0.232 | −0.101 | 1.000 | −0.443 | −0.107 |
| | Number_discussion | −0.253 | −0.261 | −0.443 | 1.000 | 0.246 |
| | Date | 0.069 | 0.577 | 0.107 | 0.246 | 1.000 |

**TABLE 3.4**
Total Variance

| | Initial Eigen Values | | | Extraction Source of Squared Loadings | | |
|---|---|---|---|---|---|---|
| **Component** | **Total** | **% of Variance** | **Cumulative %** | **Total** | **% of Variance** | **Cumulative %** |
| 1 | 1.628 | 30.327 | 30.327 | 1.628 | 30.327 | 30.327 |
| 2 | 1.151 | 21.441 | 51.769 | 1.151 | 21.441 | 51.769 |
| 3 | 1.116 | 20.789 | 72.559 | 1.116 | 20.789 | 72.559 |
| 4 | 1.06 | 19.746 | 92.306 | 1.06 | 19.746 | 92.306 |
| 5 | 0.413 | 7.693 | 100.000 | | | |

| | Components | | | |
|---|---|---|---|---|
| | 1 | 2 | 3 | 4 |
| Date | 0.194 | 0.358 | 0.906 | 0.115 |
| Specialty | 0.865 | 0.142 | −0.164 | 0.018 |
| Location | −0.294 | 0.632 | −0.270 | 0.664 |
| id | 0.869 | 0.020 | −0.136 | 0.160 |
| Number_ discussion | 0.031 | −0.776 | 0.165 | 0.601 |

**FIGURE 3.5**
Component matrix after rotation.

of more than 1. Regarding our case, four components presenting 92.30% of the total information are identified. The expert can have a hand at this stage in designating the generated components.

- *Component matrix after rotation*: Figure 3.5 indicates the correlation between variables and the four components presented in the previous step. Axis 1, named "Doctor," includes variables: Specialty and Id. In fact, these variables are most correlated with the fourth component. Axis 4, named "Topic," includes the variable "number_discussion." It is intended to illustrate the number of discussions between the doctors. Axis 2, named "Location," represents the place of a doctor. Axis 3, called "Date," represents the date of consultation and communication between doctors.

*3.3.4.2.1.1 Generation of Data Marts Schema* The logical schema of obtained data marts is presented in Figure 3.6. Indeed, it is composed of a fact related to "Topic." Among, the dimensions considered are the "Doctor" dimension; each doctor is described by their Id, Specialty, and Tel. The dimension "Location" is described by the actual place of the doctors. The dimension "Date" describes the date of discussion between the different doctors.

*3.3.4.2.1.2 Generation of "Topic" Data Marts Schema*

- *Statistical analysis*: The statistical analysis is presented in the following sections.
    - Study the correlation between variables: This step aims to reduce the data by forming a smaller number of more variables and more correlations. For this

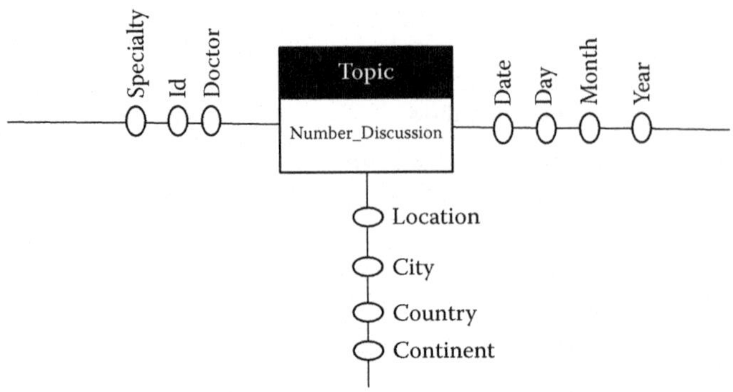


**FIGURE 3.6**
Logical schema "Topic" data marts.

end, we use the correlation matrix containing all the correlations between variables (Table 3.5). The "specialty" variable is positively correlated with the "Id" variable. "Id_P," "Disease," "Date" and "Id" variables are correlated with the "Treatment" variable. This correlation between all the variables will control the set of variables that will compose the set of factors (or components). All the correlated variables will be gathered into separate factors.

- *Extract factors using variables*: It is the total explained variance. The total variance gives us an idea about the information level represented by each component or each factor. As shown in Table 3.6, six input variables were used in order to identify five components. The first component encloses 39.07% of the total information of all variables, the second includes 62.34%, the third 80.89%, the fourth embraces 90.57%, and the fifth 98.35%. Commonly, we choose the variables that have a total superior to 1. As far as our case, four components presenting 98.35% of the total information. The expert can intervene at this stage specify the generated components.
- *Component matrix after rotation*: Figure 3.7 presents the correlation between variables and the five components identified in the previous step. Axis 1, named "Doctor," includes variables "Specialty" and "Id." In fact, these variables are most correlated with the first component. Axis 3, named "Treatment," includes the variable "TRT." They are intended to illustrate the period of treatment. Axis 5, named "Patient," represents the information of the patient. Axis 2, named "Disease." Axis 4, called "Date," represents the date of consultation and communication between doctors.

**TABLE 3.5**
Correlation Matrix

| | | Id_P | Specialty | TRT | Disease | Date | Id |
|---|---|---|---|---|---|---|---|
| Correlation | Id_P | 1.000 | 0.690 | 0.721 | 0.037 | 0.920 | 0.902 |
| | Specialty | 0.690 | 1.000 | 0.948 | −0.207 | 0.715 | 0.690 |
| | TRT | 0.721 | 0.948 | 1.000 | −0.140 | 0.774 | 0.721 |
| | Disease | 0.037 | −0.207 | −0.140 | 1.000 | 0.085 | 0.037 |
| | Date | 0.920 | 0.715 | 0.774 | 0.085 | 1.000 | 0.920 |
| | Id | 0.902 | 0.690 | 0.721 | 0.037 | 0.920 | 1.000 |

**TABLE 3.6**
Total Variance

| | Initial Eigen Values | | | Extraction Source of Squared Loadings | | |
|---|---|---|---|---|---|---|
| **Component** | **Total** | **% of Variance** | **Cumulative %** | **Total** | **% of Variance** | **Cumulative %** |
| 1 | 5.245 | 39.079 | 39.079 | 5.245 | 39.079 | 39.079 |
| 2 | 3.122 | 23.261 | 62.340 | 3.122 | 23.261 | 62.340 |
| 3 | 2.491 | 18.557 | 80.897 | 2.491 | 18.557 | 80.897 |
| 4 | 1.299 | 9.682 | 90.579 | 1.299 | 9.682 | 90.579 |
| 5 | 1.043 | 7.775 | 98.353 | 1.043 | 7.775 | 98.353 |
| 6 | 0.222 | 1.647 | 100.00 | | | |

| | Components | | | | |
|---|---|---|---|---|---|
| | 1 | 2 | 3 | 4 | 5 |
| Id_P | 0.143 | 0.167 | −0.267 | 0.106 | −0.319 |
| Specialty | 0.875 | −0.287 | 0.155 | 0.093 | 0.132 |
| TRT | 0.201 | −0.208 | 0.577 | −0.042 | −0.148 |
| Disease | −0.049 | 0.953 | 0.297 | 0.030 | 0.003 |
| Date | 0.341 | 0.181 | −0.113 | −0.256 | 0.057 |
| Id | 0.943 | 0.167 | −0.267 | 0.106 | −0.019 |

**FIGURE 3.7**
Component matrix after rotation.

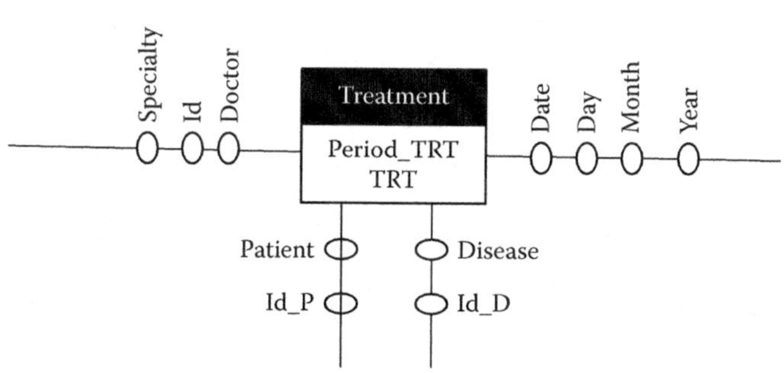


**FIGURE 3.8**
Logical schema "Treatment" data marts.

*3.3.4.2.1.3 Generation of "Treatment" Data Marts Schema* The logical schema of obtained data marts is presented in Figure 3.8. Indeed, it is composed of a fact related to "Treatment." Among, the dimensions considered is the "Doctor" dimension, each doctor is described by their Id and Specialty. The dimension "Location" is described by the actual place of the doctors. The dimension "Date" describes the date of discussion between the different doctors. Finally, the "Disease" dimensions "Patient."

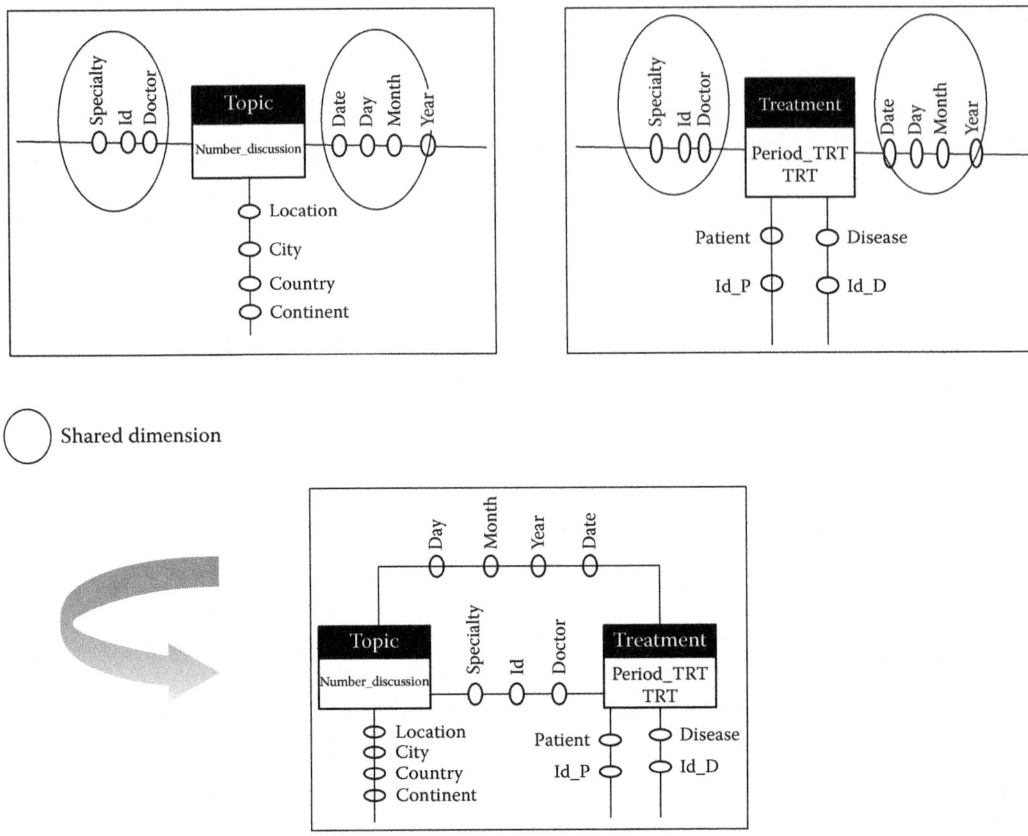


**FIGURE 3.9**
Constellation schema data warehouse.

### 3.3.4.3 *Step 3: Generation of Data Warehouse Schema*

The constellation schema of obtained data warehouse is presented in Figure 3.9. Indeed, it is composed of two facts related to "Topic" and "Treatment." We note the existence of two shared dimensions between the two data marts schema that have been previously obtained, that is, "Doctor" and "Date." The new data warehouse schema is obtained by joining these two patterns. The result of schema generation is then validated by the domain expert.

## 3.4 Conclusion

Data stores are a cheaper database as the data warehouse, and lighter for some users, as a department. This is a small and focused structure and driven by user needs. It is interesting to start with the design of data stores, provided you follow certain rules. In this chapter, we proposed a new approach to design and construction of data warehouse based on the basics of descriptive statistics for the analysis of multidimensional data PCA.

## References


1. Inmon, W.H. *Building the Data Warehouse*. John Wiley & Sons, Indianapolis, IN, 1996.
2. Hüsemann, B., Lechtenbörger, J., and Vossen, G. Conceptual data warehouse design. In *Proceedings of the International Workshop on Design and Management of Data Warehouses*. Stockholm, Sweden, pp. 3–9, 2000.
3. Golfarelli, M., Maio, D., and Rizzi, S. The dimensional fact model: Conceptual model for data warehouses. *International Journal of Cooperative Information Systems*, 7, 215–247, 1998.
4. Romero, O. and Abelló, A. Automating multidimensional design from ontologies. *DOLAP'07*, Lisboa, Portugal, November 9, 2007.
5. Luján-Mora, S., Trujillo, J., and Song, I.Y. Extending the UML for multidimensional modeling. *Proceedings of the International Conference on the Unified Modeling Language*, Dresden, Germany, pp. 290–304, 2002.
6. Luján–Mora, S., Trujillo, J., and Song, I.Y. A UML profile for multidimensional modeling in data warehouse. *Data and Knowledge Engineering*, 59(3), 725–769, 2006.
7. Rizzi, S. Conceptual modeling solutions for the data warehouse. *Database Technologies: Concepts, Methodologies, Tools, and Applications*. pp. 86–104, 2009.
8. Mazón, J., Trujillo, J., Serrano, M., and Piattini, M. Designing data warehouses: From business requirement analysis to multidimensional modeling. *REBNITA Requirements Engineering for Business Needs and IT Alignment*, Cox, K., Dubois, E., Pigneur, Y., Bleistein, S.J., Verner, J., Davis, A.M., and Wieringa, R. (eds.). University of New South Wales Press, Sydney, New South Wales, Australia, 2005.
9. Cabibbo, L. and Torlone, R. A logical approach to multidimensional databases. In *International Conference on Extending Database Technology (EDBT 98)*, Valencia, Spain, LNCS, Springer, pp. 183–197, 1998.
10. Giorgini, P., Rizzi, S., and Garzetti, M. Goal-oriented requirement analysis for data warehouse design. *Proceedings of Eighth International Workshop on Data Warehousing and OLAP*, ACM Press, pp. 47–56, DOLAP 2005.
11. Giorgini, P., Rizzi, S., and Garzetti, M. A goal-oriented approach to requirement analysis in data warehouses. *Decision Support Systems (DSS) Journal*, 45(1), 4–21, 2008, Elsevier.
12. Vassiliadis, P., Simitsis, A., and Skiadopoulos, S. Conceptual modeling for ETL processes. Theodoratos, D. (ed.), *DOLAP 2002, Proceedings of the 5th ACM international workshop on Data Warehousing and OLAP*, McLean, Virginia, November 08, 2002, pp. 14–21, 2002.
13. Kimball, R. *The Data Warehouse Toolkit*. John Wiley & Sons, Inc., New York, 1996.
14. Bonifati, A., Cattaneo, F., Ceri, S., Fuggetta, A., and Paraboschi, S. Designing data marts for data warehouses. *ACM Transactions on Software Engineering and Methodology*, 10, 452–483, 2001.
15. Nabli, A., Feki, J., and Gargouri, F. Automatic construction of multidimensional schema from OLAP requirements. *Arab International Conference on Computer Systems and Applications (AICCSA'05)*, Cairo, Egypt, IEEE, January 2005.
16. Tebourski, W., Karra, W., and Ben Ghezala, H. Semi-automatic data warehouse design methodologies: A survey. *IJCSI International Journal of Computer Science Issues*, 10(5), 2, September 2013.
17. Boufaïda, Z., Yahiaoui, L., and Prié, Y. Semantic annotation of documents applied to E-recruitment. *SWAP, The Third Italian Semantic Web Workshop*, Pisa, Italy, pp. 1–6, 2006.
18. Jr. Hair, J.F., Black, C., Babin, W., Anderson, R.E., and Tatham, R.L. *Multivariate Data Analysis*, 5th edn. Pearson-Prentice Hall, Upper Saddle River, NJ, 2006.

# 4

# *Multi-Agent System for Text Mining*

**Safa Selmi and Wahiba Ben Abdessalem Karaa**

**CONTENTS**

**ABSTRACT** Nowadays, the amount of textual information has become increasingly important. Text mining concepts and techniques are applied to extract significant information from a text. This valuable knowledge can be extracted and subsequently returned to the user. In order to achieve this target, a lot of researches have been proposed. Many techniques have been discovered, but the results are different in terms of efficiency. To improve this factor, a multiagent system is one of the solutions. In this chapter, we detail text mining and multiagent systems and to show the necessity of this technology in order to find the adequate information for the user.

**KEY WORDS:** *text mining, multi-agent systems.*

## 4.1 Introduction

Following the globalization of commerce and the development of the Internet, the production of e-documents is soaring. As a result, researchers have discovered new solutions in order to produce, diffuse, research, exploit, and translate information into text. The concept of text mining has already been introduced in a wide number of scientific publications. Text mining is a specific step in the general process of knowledge extraction through texts.

Many types of research have been undertaken using the multi-agent systems (MAS) paradigm in the field of text mining. The main objective of the theory behind the MAS is based on calculating the distributed artificial intelligence (DAI). Distributed artificial intelligence (DAI) is a predecessor of the field of multiagent systems. Indeed, there are many techniques to solve distributed problems where a group of agents can dynamically discuss: how to spread a problem, how to distribute the different subtasks, and how to exchange information and solve the possible independence between the partial solutions. MAS is the latest generation of intelligent software and their primary target is to apprehend, simulate, model, and solve the heterogeneous, complicated, and evaluative systems via interactions between autonomous entities called agents. In text mining, the agents can work simultaneously and interact to process, analyze, search, and extract information or knowledge efficiently.

The remaining sections of this chapter are organized as follows. In Section 4.2, natural language processing is detailed. In Section 4.3, text mining is presented. The paradigm of multi-agent systems will be introduced in Section 4.4. Section 4.5 will discover the different searches between the multi-agents systems and text mining. Section 4.6 reports the conclusion and future perspectives.

## 4.2 Natural Language Processing

### 4.2.1 NLP Definition

Natural language processing (NLP) is a field in computer science and linguistics that is closely related to Artificial Intelligence and Computational Linguistic. It can be defined as follows: "Natural Language Processing is a theoretically motivated range of computational techniques for analyzing and representing naturally occurring texts at one or more levels of linguistic analysis for the purpose of achieving human-like language processing for a range of tasks or applications" [1].

Hence, NLP is concerned with the use of computational mechanisms to perform communication between a person and a computer through human language. Thus, NLP can be used in many applications. For example, users' requirements are expressed in natural language. They have to be amenable to NLP in order to derive valuable information [2–5].

### 4.2.2 NLP Applications

The most frequent applications using NLP include the following:

*Machine translation* refers to the automated translation of text from one human language to another assisted by computer [6].

*Information retrieval* (IR) is generally concerned with the representation, storage, organization, and access to information items such as text documents, sound, images, or data [6].

*Information extraction* (IE) is the process of deriving, from digital text documents written in natural language, structured information that expresses relationships between entities and transforming them into a structured representation (e.g., a database) [6].

*Automatic summarization* is the creation of a shortened version of a text by means of a computer program. The generated document contains the most important points of the original document [6].

*Speech recognition* is a computer-driven conversion of a speech signal (i.e., voice) into readable text [6].

### 4.2.3 NLP Levels

Generally, the processing stages in an NLP system are morphology level, lexical level, syntactic level, semantic level, discourse level, and pragmatic level [7].

*Morphology level* is the process in which a word is analyzed into its root word and associated morphemes [7]. For example, the word "preregistration" can be morphologically analyzed into three separate morphemes as shown in Figure 4.1.

*Lexical level* is about breaking up the input text into basic units of the source language called tokens (e.g., words, punctuation marks, and discards whitespace). Therefore, this process is called "tokenization." A text could be broken up into paragraphs, sentences, words, syllables, or phonemes [7].

*Syntactic level* deals with the construction of sentences. It indicates how the words are related to each other, so as to uncover the grammatical structure of the sentence [7].

*Semantic level* produces a representation of the meaning of the text [7].

*Discourse level* focuses on the effect of the previous sentence on the next sentence [7].

*Pragmatic level* interprets a sentence according to its meaning, which requires good world knowledge, including the understanding of intentions and goals [7].

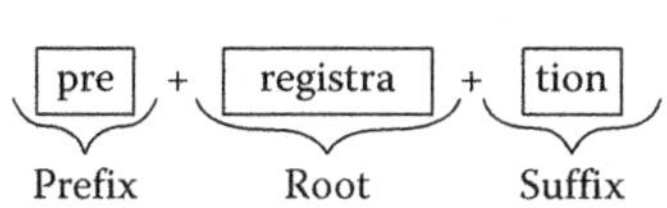


**FIGURE 4.1**
Example of morphological analysis.

Morphological, lexical, syntactical, and semantic levels deal with words, phrases, and sentences, while discourse and pragmatic levels deal with whole paragraphs and dialogues. Each level of the text analysis process includes several tasks to be carried out.

## 4.3 Text Mining

Text data mining is increasingly becoming more vital these days, as the ability to extract high-value information remains limited. This engenders a difficult problem with its automatic processing in order to meet the needs of users who are searching for relevant information. Methods of text mining (TM) and natural language processing (NLP) can partly solve such a challenge.

In fact, they involve modeling and implementing methodologies applied to textual data in order to determine the direction, or discover new information.

The process of text mining is composed of two consecutive phases: text refining and knowledge distillation (Figure 4.2).

The first phase consists of transforming of free-form text documents into a chosen intermediate form. The second phase allows extracting patterns or knowledge from intermediate forms, such as the most significant keywords. These words are to be extracted and then to be treated in order to summarize the content of the texts. The descriptors of language can be simple words (e.g., "war"), but also compound terms (e.g., "family agriculture"). The intermediate form (IF) can be document based, wherein each entity represents an object or concept of interests in a specific domain. Moreover, mining a document-based IF takes the different patterns and relationships between documents [8].

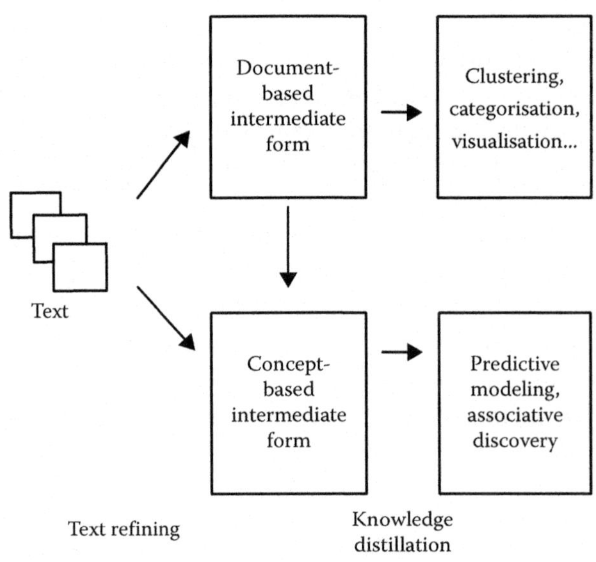


**FIGURE 4.2**
General framework for text mining.

### 4.3.1 A General Definition of Extracting Information from Texts

Information management tools need to extract existing terminology in texts to meet the user requirements of trying to find relevant information. However, the acquisition or extractions of these terminologies are still a critical issue. Therefore, it is necessary to have automatic or semiautomatic extraction terminology systems from texts. These tools use three approaches: linguistic approach, statistical approach, and hybrid approach.

### 4.3.2 Linguistic Approaches

These approaches are basically used for technique linguistic analysis, which is based on the knowledge of language and its structure. These approaches exploit syntactic knowledge, lexical or morphological.

#### *4.3.2.1 TERMINO*

The TERMINO tool is the first tool that is used for automatic extraction terms. This system was developed as part of a collaboration between a team of the ATO center of the University of Quebec-Montreal and the office of the French language in Quebec in the late 1980s [9]. The modern form of TERMINO is NOMINO software that extracts candidate terms (called synapses) through the identification of noun phrases in the corpus. The TERMINO processing chain consists of three stages [10]:

1. *Preprocessing of the text*: In this stage, the text is divided into tokens and then filtered to remove the formatting characters. This step is necessary for any terminology extraction process from the textual corpus.
2. *Lemmatization lexemes*: In this step, each identified token is subject to morpho-syntactic analysis to assign a grammatical category to each token.
3. *Disambiguation*: This step is useful to perform a syntactic analysis in context to disambiguate the tokens that have more than one grammatical category late lemmatization step. After this step, all the tokens of the text have only one grammatical category.

#### *4.3.2.2 LEXTER*

LEXTER was developed by D. Bourigault [11,12]. LEXTER is a Terminology Extraction Software. LEXTER performs a morpho- syntactical analysis of a corpus on any technical subject is fed in it.

It's completely different to TERMINO, the corpus treated LEXTER are tagged and disambiguated [13]. To extract candidate terms, LEXTER performs morpho-syntactic analysis that allows it to identify and analyze noun phrases to term. The result is a set of candidate terms organized into a grammatical network. The acquisitions of candidate terms involve several steps:

A morphological analysis step assigns to the words of the sentence grammatical label; punctuation is also labeled. The maximum nominal groups are identified by making their syntactic boundaries (conjugated verbs, pronouns, conjunction, etc.).

A decomposition step analyzes recursively maximal noun phrases. The maximum nominal phrases and components are returned as candidate terms.

A structuring step maps the candidate terms in a terminology network based on the decomposition of these terms. In this network, each term is connected to its head and expansion, and each head and each expansion are connected to the compound terms to which they belong. This approach uses endogenous learning techniques to solve the prepositional attachment ambiguities and problems in the adjectival noun.

The current version of LEXTER [11] is a syntactic analysis in French or in English. This tool sends, as a result, a group of words and phrases. Also, it was adopted in a large number of searches for different subjects. For example, in [12] it is used in an ontology construction method via the domain body of chirurgical reanimation.

#### *4.3.2.3 SYNTEX*

SYNTEX identifies a group of nouns, verbs, adjectives, and nominal, verbal, and adjectival phrases. These elements form a network of syntaxes dependency also called conceptual network by building links between every candidate term and other candidate/chosen terms in which it has the head or expanded position. The components of the network form together with the system or the group of candidate terms. Then SYNTAX has, respectively, two roles: for each candidate term, it provides its frequency in the corpus and its productivity on head or on expansion position of other candidate terms.

Thus, the linguistic approaches find their performances in very specific corpuses in which a linguistic detailed study has been done. These approaches cannot be generalized among the corpuses of a different language, different size, and different specialty.

### 4.3.3 Statistical Approaches

Essentially, statistical approaches are based on quantitative techniques. These approaches are often used for the treatment of very large corpuses. Due to the development of new technologies, digital documents are becoming easily available and facilitate the formation of these large corpuses.

The ANA (natural automatic acquisition) system [13,14] is a terminology extraction tool. Accordingly, ANA is mainly based on statistical technique systems. The system accepts to process new data, and it takes place in two phases.

In the first module, which is called "familiarization," the software extracts knowledge in corpus form of four lists, separating the function words of candidate terms (conjunction, adverbs, etc.). This list of candidate terms is then enriched in the second module, "discovery," based on co-occurrences identified in the corpus [14,15].

The candidate terms extracted by ANA are presented with a list of candidate terms with their variants identified in the corpus or couples "candidate terms/frequency in the corpus" or by semantic network.

The recognized benefits of statistical approaches are essentially the ability to process large size of corpus and independence of linguistic resources (such as dictionaries or grammars) external to the treaty body. The latter makes it faster and also more economical because the linguistic resources are often expensive.

### 4.3.4 Hybrid Approaches

In hybrid approaches, statistical and linguistic approaches are combined in such order in which this association varies from one system to another. Indeed, in some systems, the results obtained by a linguistic analysis are validated and filtered by a statistical analysis, while in other systems the results of the statistical analysis are validated by a linguistic analysis.

#### 4.3.4.1 ACABIT

The system ACABIT was developed by B. Daille [16–18] from IBM company, which works on a previously tagged and disambiguated corpus. It is only dedicated to the automatic extraction of candidate terms, by parsing the corpus, followed by statistical processing to filtering the results of the analysis. In the first step, the system uses the technical language used by TERMINO and LEXTER [19]. Also, ACABIT based on a important corpus of terms, and it performs a syntactic analysis followed by statistical processing. ACABIT begins by collecting the syntactic schemas of simple terms (N,N 0 D et N, etc.), and mechanisms of variation make it possible to obtain more complex terms like an automation. According to Dunning [20], ACABIT is based on the use of various statistical measures that retain the best candidate terms without being sensitive to frequencies. The system is based on a body of reference and a valid list of terms. According to Dunning [20], statistical measurements seem to be the best to represent the candidate terms.

#### 4.3.4.2 XTRACT

The XTRACT tool was developed by Smadja [21] through his work on the automatic indexing of texts. It involves identifying collections of predefined structures such as name+name, name of name, name+adjective, subject+verb, verb+subject, and so on. Initially, XTRACT essentially uses statistical techniques based on the mutual information between words, and linguistic techniques thereafter.

We emphasize that there is not much difference between the XTRACT and ACABIT systems. In XTRACT, the results obtained by statistical methods are subject to filtering by language techniques. In ACABIT, we find the opposite path by performing statistical techniques filtering the results of linguistic techniques.

#### 4.3.4.3 TERMS

The extracted segments are filtered according to their frequency in the corpus in order to eliminate nonrepeated segments. The terms' extraction is performed by the recognition of syntactic patterns from a tagged corpus. The patterns are written from the study of the syntactic construction of terminological dictionary entries. The extracted segments are then filtered according to their frequency in the corpus in order to eliminate nonrepeated segments.

This hybrid approach takes advantage of the speed and independence from the field of statistical methods. This independence is manifested by the lack of use of specialized language resources and dictionaries. However, this independence remains very partial and limited, because the linguistic methods should require a perfect knowledge of the language of the corpus to be treated.

## 4.4 Multi-Agent Systems

The emerging of technology agent has invited a lot of attention in recent years because of its great potential and the emergence of new technology systems based on agents. Nowadays, it is recognized as a promising technology for the construction of autonomous, complex, and intelligent systems [23].

The theory behind the multi-agent systems is based on the calculation of distributed artificial intelligence. These are considered the latest generation of intelligent software systems and their primary objective is to understand, simulate, model, and solve the heterogeneous systems.

### 4.4.1 Definition of a Multi-Agent System

A multi-agent system (MAS) is an organized set of agents; it consists of one or more organizations that shape the rules of the coexistence and the collective work of these agents.

The multi-agent system is used to reduce the complexity of a problem by decomposing it into simpler subsystems; each subsystem is assigned to an agent to ensure the coordination of other tasks with other agents [24].

A multi-agent system or SMS is a distribution system composed of set agents that interact to cooperate, coexist, and compete [25].

#### *4.4.1.1 The Benefits and the Reasons for Using a Multi-Agent System*

The benefits of the use of multi-agent systems are the following [26]:

- Increase of efficiency and speed, because of parallel calculus
- Low cost
- Easy development and reuse, because it is easier to create and maintain a system that is composed of modules
- Scalability and flexibility to add new agents

The multi-agent solution is recommended for these reasons [27]

- Integration of distributed data sources due to the communication and the organization of agents
- Ability to solve problems that are too difficult to resolve by centralization of systems
- Simulations of problems when the objects act independently
- Knowledge integration of distribution domain

### 4.4.2 Definitions of an Agent

In the literature there are several definitions for an agent; one of the most famous definitions was presented by Ferber [24]: "an agent is a computer entity situated in an environment and able to act in this environment, to receive a limited way and to represent partially this environment, to communicate with other agents, follow internal trends (goals, satisfaction research…)".

According to Russel and Norvig [28], the structure of an agent can be divided into three parts:

1. *The programming part of agent*: The part that implements a mapping perception actions (which is called function selection of actions).
2. *The status of the agent*: It includes all internal representations over the agent program operations. This may include the representation of the agent's environment and objectives.

3. *The architecture of agent*: It is a virtual machine that makes perceptions from the sensors of the agent available for the agent program.

There are four characteristics that distinguish the agents and allow them to perform actions in an autonomous way: autonomy, responsiveness, proactivity, and social relations.

An agent has a collective behavior, consequence of perceptions, representations, and interactions with the environment and other agents [29]. On the basis of the criterion of intelligence, agents can be classified into three types [30]:

1. Reactive
2. Cognitive
3. Hybrid

### 4.4.3 Types of Agents

#### *4.4.3.1 The Reactive Agents*

The reactive agents are the most basic agents; they have a behavior type's stimulus response [28]. Indeed, intelligent behaviors can emerge from their association. It's capable to respond in a timely manner to changes in its environment.

A reactive agent is typically composed of four parts:

1. Control
2. Behavior
3. Perception
4. Reproduction

#### *4.4.3.2 The Cognitive Agents*

The cognitive agents are more evolved; they possess a global representation of their environment and the other agents with whom they communicate; they know how to take into account their past and get organized around a social model of organization.

The cognitive agents can be distinguished under three categories based on the main characteristics of the artificial intelligence, namely, degrees of autonomy, cooperation, and adaptation.

The autonomy is reflected by the ability of agents to take initiatives to achieve their goals. For the cooperation between the agents, it is necessary to constitute a coherent system.

Regarding the adaptation, the agent must adapt the dynamic environment to achieve its goal.

#### *4.4.3.3 Cognitive Agents versus Reactive Agents*

The cognitive agents are more evolved. These agents have reasoning abilities through individual (e.g., deduction capabilities) rather than global behavior performance results of their interactions.

These agents have the reasoning abilities through individual (e.g., deduction capabilities) rather than global behavior performance results of their interactions.

Cognitive agents are intelligent agents and are capable of solving complex problems; these agents are specific to the environment and agent skills. However, their behavior is complex because it is linked to their goals.

Cognitive agents are also known as intentional agents because they have objectives and plans to accomplish their explicit goals [31]. The reactive agents are perceived as a biological approach to the concept of an agent.

An agent is said to be a reactive agent if it answers in a timely manner to changes in the environment. This type of agent has no memory of its history and global goal. For this reason, a reactive multi-agent system is composed of a large number of agents that despite their simple structure can be capable of complex group and coordinated action.

#### *4.4.3.4 Hybrid Agent*

The principal idea is to structure the features of this agent in two or more hierarchical layers based on the types of conventional agents that interact with it in order to achieve a consistent state.

These agents have the following advantages:

- The capabilities of this agent will increase because the various layers can be executed in parallel.
- The reactivity of the agent is improved because the agent can reason in a symbolic world while supervising their environment and reacting in sequence.

## 4.5 Multi-Agent System for Text Mining

A multi-agent system is a powerful tool in the development of complex systems. Using multi-agent systems has a number of advantages like autonomy, social proactivity, and reactivity. This solution has become one of the more optimistic to solve this problem [32].

In the recent years, the necessity of this approach has increased because it is widely used in computer and it can be a very optimistic solution to this problem [32].

Many approaches have been proposed in the domain of MAS and text mining. Indeed, there is a strong demand for web text mining, which helps people to find information and also to discover some useful information from web documents. Therefore, the MAS paradigm is used in order to meet the requirements in a specific time.

A search engine is a web-based tool that enables users to locate information on the World Wide Web. More than 90% of the information can be retrieved from the search engine by users. Search engines utilize automated software applications (referred to as robots, bots, or spiders) that travel along the web, following links from page to page, site to site. The information gathered by the spiders is used to create a searchable index of the web.

A multi-agent system is basically used to personalize information, which helps to filter and search information.

In this context, Widyantoro et al. [33] introduced a large number of representations for user profiles in order to personalize information systems. They focused on the importance of long- and short-term user profiles usage. They introduced a technique that records user behavior and presents it with "implicit feedback," generally referring to data traits that a customer leaves behind them when naturally interacting with a site or platform.

Nick and Themis [34] used the user feedback in order to build an MAS that enables to realize metasearch. In addition, through this system, the user can analyze their own model and indicate to the system what the important terms are. It reduces work overload by using explicit feedback. Bottraud et al. [35] introduced the mechanism or automated heuristic for the acquisition of implicit feedback, for example, extract concepts from the work context. Enembreck [36] proposed the use of personal documents from a clustering algorithm, while Bottraud et al. [35] used a structure of standard vector to present the center of interest of the user.

Enembreck [36] also proposed an approach *MAIS* (Multi-Agent-based Internet Search), using an open multi-agent system containing personal assistants, library agents, filter agents, and search agents. This intelligent system is considered as a model of Internet multi-agent systems. The types of messages exchanged between these agents are Request, Answer, and Call-For-Bids (CFB). For the dynamic allocation of tasks, documents containing vector model, documents, or graphs of concepts are used by the algorithm to learn how to analyze the level of relevance of a document. This multi-agent system is an open system. Indeed, it is available to add new agents during the system execution, without affecting the global functions. This model has many advantages:

- Aptitude to define the agents playing the metasearch role and able to retrieve pages provided by several existing search engines like Google, AltaVista, or All-the-Web.
- Aptitude of agents to build and update profiles of users. Thus, people who have different interest centers can receive different results for the same query. Any user (even those who do not have much experience) can profit by some customization.
- Aptitude of *MAIS* to retrieve and to classify, automatically, the information from research done by the standard search engines.

But, these tasks are very specific and are not conducive for the resolution of distributive problems, because the activities of each agent are executed independently.

The use of MAS is involved in many fields. For instance, Lai [37] implemented a multi-agent web text mining system to support company decision making. It used a framework of the back propagation neural network (BPNN) based on intelligent learning agent for the text mining. It refers to the process of using unstructured web-type text and examining it to discover implicit patterns hidden within the web documents. The BPNN is used as an intelligent agent to explore the hidden patterns. It is a supervisor learning mechanism in the form of the neural network association memory. But a single BPNN agent cannot handle large-scale text documents. With the rapid increase of web information, a multi-agent web text mining system on the grid is constructed for large-scale text mining applications. The multi-agent-based web text mining on the grid can discover some useful knowledge for enterprise decision support in an efficient way. But, this architecture is currently under implementation and the tests will be performed in simulation.

MAS are also applied to information retrieval (IR) to obtain relevant information resources within large collections. In this context, Lee [38] proposed a simple system for information retrieval agents based on many terms or keywords' distribution in a document. The techniques related to meaningful terms' frequency and keyword distribution characteristics are used in this extraction model. In this model, terms are selected by using stemming and filtering stop lists. The agents receive information from web clients and extract key paragraph frequency and keywords, and then the agents construct the profile of the documents with the keywords, key paragraph, and address of the document.

Cao et al. [39] have described the summarized view of agent-mining interaction from the perspective of an emerging area in the scientific family. It includes vital driving forces, originality, primary research directions, specific issues, development of research groups, declarations, and movement of agent-mining interaction. It indicates both theoretical and application-oriented features. It analyzes the research directions and interactive improvement matters in agent–mining interaction and approach to this relation.

## 4.6 Conclusion and Perspective

In this chapter, the related work of multi-agent systems and text mining is discussed. We notice that the majority of the works combining text mining and multi-agent systems are used in the context of the web. Our future work is to propose an approach for automatic extraction of concepts and relationships between these concepts in the context of MEDLINE biomedical bibliographic database.

## References

1. Liddy, E. D. (2001). Natural language processing. *Encyclopedia of Library and Information Science*, Marcel Decker, Inc.
2. Karaa, W. B. A., Ben Azzouz, Z., Singh, A., Dey, N., Ashour, S. A., Ben Ghazala, H. (2015). Automatic builder of class diagram (ABCD): An application of UML generation from functional requirements. *Software: Practice and Experience.*
3. Abdouli, M., Karaa, W. B. A., and Ghezala, H. B. (June 2016). Survey of works that transform requirements into UML diagrams. *2016 IEEE 14th International Conference on Software Engineering Research, Management and Applications (SERA)* (pp. 117–123). IEEE.
4. Herchi, H. and Abdessalem, W. B. (2012). From user requirements to UML class diagram. *arXiv preprint arXiv:1211.0713.*
5. Joshi, S. D. and Deshpande, D. (2012). Textual requirement analysis for UML diagram extraction by using NLP. *International Journal of Computer Applications*, 50(8), 42–46.
6. Liddy, E. D., Hovy, E., Lin, J., Prager, J., Radev, D., Vanderwende, L., and Weischedel, R. (2003). Natural language processing. *Encyclopedia of Library and Information Science*, 2.
7. Indurkhya, N. and Damerau, F. J. (Eds.). (2010). *Handbook of Natural Language Processing* (Vol. 2). CRC Press, Boca Raton, FL.
8. Sumathy, K. L. and Chidambaram, M. (October 2013). Text mining: Concepts, applications, tools and issues: An overview. *International Journal of Computer Applications (0975–8887)*, 80(4).

9. Benveniste, E. (1966). Formesnouvelles de la composition nominale. *Bulletin de la Société de linguistique*, de Paris, LX1 (1), 82–95. Republished, Problèmes de linguistique générale, 2, Gallimard, Paris, (1974).
10. Bourigault, D. (August 1992). Surface grammatical analysis for the extraction of terminological-noun phrases. *Proceedings of the 14th Conference on Computational Linguistics* (Vol. 3, pp. 977–981). Association for Computational Linguistics.
11. Bourigault, D. (1994). Lexter: unLogicield'EXtraction de TERminologie: application à l'acquisition des connaissances à partir de textes. Doctoral dissertation, EHESS, Paris, France.
12. Bourigault, D., Gonzalez-Mullier, I., and Gros, C. (August 1996). LEXTER, a Natural Language Processing tool for terminology extraction. *Proceedings of the Seventh EURALEX International Congress* (pp. 771–779).
13. Le Moigno, S., Charlet, J., Bourigault, D., and Jaulent, M. C. (2002). Construction d'uneontologie à partir de corpus: Expérimentationet validation dans le domaine de la réanimationchirurgicale. *Actes des*, 6, 229–238.
14. Enguehard-Gueiffier, C. (1992). ANA: Acquisition NaturelleAutomatique d'un réseausémantique. Doctoral dissertation, Compiègne, France.
15. Enguehard, C. (1993). Acquisition de terminologie à partir de gros corpus. *Informatique & Langue Naturelle, ILN*, 93, 373–384.
16. Daille, B. (1994). Approchemixte pour l'extraction de terminologie: statistiquelexicaleetfiltreslinguistiques. Doctoral dissertation.
17. Daille, B. (1996). Study and implementation of combined techniques for automatic extraction of terminology. *The Balancing Act: Combining Symbolic and Statistical Approaches to Language*, 1, 49–66.
18. Daille, B. (1999). Identification des adjectifsrelationnels en corpus. *Actes de TALN*, 105–114.
19. David, S. and Plante, P. 1990. De la nécessitéd'uneapprochemorphosyntaxiquedansl'analyse de textes. *Intelligence artificielle et sciences cognitives au Québec*, 3(3), 140–154.
20. Dunning, T. (1993). Accurate methods for the statistics of surprise and coincidence. *Computational Linguistics*, 19(1), 61–74.
21. Smadja, F. (1993). Retrieving collocations from text: Xtract. *Computational Linguistics*, 19(1), 143–177.
22. Séguéla, P. (2001). Construction de modèles de connaissances par analyselinguistique de relations lexicalesdans les documents techniques. *Mémoire de thèse en Informatique, Université Toulouse*, 3, TAL, volume 47 – n° 1/2006, pp 11 à 32.
23. Hakansson, A., Thanh Nguyen, N., Hartung, R., Howlett, R. J., and Jain, L. C. (2010). Conference report of the third *KES Symposium on Agent and Multi-Agent Systems: Technologies and Applications. International Journal of Knowledge-Based and Intelligent Engineering Systems*, IOS Press, 14, 45–47.
24. Ferber, J. and Perrot, J. F. (1995). *Les systèmes multi-agents: versune intelligence collective.* InterEditions, Paris.
25. Vlassis, N. (2007). A concise introduction to multiagent systems and distributed artificial intelligence. *Synthesis Lectures on Artificial Intelligence and Machine Learning*, 1(1), 1–71.
26. Weiss, G. (1999). *MultiagentSystems: A Modern Approach to Distributed Artificial Intelligence.* MIT Press, Cambridge, MA.
27. Pipattanasomporn, M., Feroze, H., and Rahman, S. (March 2009). Multi-agent systems in a distributed smart grid: Design and implementation. *IEEE/PES Power Systems Conference and Exposition, 2009 PSCE'09* (pp. 1–8). IEEE.
28. Russell, S. and Norvig, P. (1995). *Artificial Intelligence: A Modem Approach.* Prentice Hall, Upper Saddle River, NJ.
29. Bonabeau, E. (2002). Agent-based modeling: Methods and techniques for simulating human systems. *Proceedings of the National Academy of Sciences*, 99(suppl. 3), 7280–7287.
30. Müller, J. P. (2002). Des systèmesautonomes aux systèmes multi-agents: Interaction, émergenceetsystèmes complexes. Doctoral dissertation, UniversitéLibre de Bruxelles, Brussels, Belgium.

31. Adam, C., Gaudou, B., Hickmott, S., and Scerri, D. (2011). Agents BDI et simulations sociales. *Revue d'IntelligenceArtificielle (RIA)-Num. Spec. Simul. Multi-Agent*, 25(1), 11–42.
32. Roche, R., Blunier, B., Miraoui, A., Hilaire, V., and Koukam, A. (November 2010). Multi-agent systems for grid energy management: A short review. *IECON 2010–36th Annual Conference on IEEE Industrial Electronics Society* (pp. 3341–3346). IEEE.
33. Widyantoro, D. H., Ioerger, T. R., and Yen, J. (2001). Learning user interest dynamics with a three-descriptor representation. *Journal of the American Society for Information Science and Technology*, 52(3), 212–225.
34. Nick, Z. Z. and Themis, P. (2001). Web search using a genetic algorithm. *IEEE Internet Computing*, 5(2), 18.
35. Bottraud, J. C., Bisson, G., and Bruandet, M. F. (July 2003). Apprentissage de profilspour un agent de recherched'information. *Actes de la Conférence Apprentissage (CAP 2003)* (pp. 31–46).
36. Enembreck, F. (2003). Contribution à la conception d'agentsassistantspersonnelsadaptatifs. Doctoral dissertation, Compiègne, France.
37. Lai, K. K., Yu, L., and Wang, S. (January 2006). Multi-agent web text mining on the grid for enterprise decision support. *Asia-Pacific Web Conference* (pp. 540–544). Springer, Berlin, Germany.
38. Lee, J. W. (April 2007). A model for information retrieval agent system based on keywords distribution. *Proceedings of the 2007 International Conference on Multimedia and Ubiquitous Engineering* (pp. 413–418). IEEE Computer Society.
39. Cao, L., Luo, C., and Zhang, C. (June 2007). Agent-mining interaction: An emerging area. *International Workshop on Autonomous Intelligent Systems: Multi-Agents and Data Mining* (pp. 60–73). Springer, Berlin, Germany.

# 5

# *Transformation of User Requirements in UML Diagrams: An Overview*

**Mariem Abdouli, Wahiba Ben Abdessalem Karaa, and Henda Ben Ghezela**

**CONTENTS**



**ABSTRACT** This chapter focuses on the process of extraction of UML (*U*nified *M*odeling *L*anguage) diagrams from requirements written in natural language. Indeed, this process is in the context of requirement engineering (*RE*). The goal of RE is to translate the objectives of the real world into a system that complies with the requirements. However, several factors cause erosion during this process, such as delay, linguistic errors, meaning conflicts, and so on. These erosions are due to manual tasks, approaches that do not provide the desired result. This chapter provides a survey on the transformation of requirements into UML diagrams and a comparison between existing approaches.

**KEY WORDS:** *requirement, transformation, UML diagrams.*

## 5.1 Introduction

Nowadays, the modeling phase in software development is separated from other phases by developers, because the process is difficult and critical. The separation of phases is done by the use of model-driven engineering (MDE).

MDE is an approach based on modeling. The increasing complexity in software development was the reason for the emergence of MDE, to overcome difficulties.

To create and transform models or UML diagrams, software development incorporates several steps. The creation and the transformation of models via MDE are provided by languages, concepts, and tools. The benefits of using MDE is to cover the whole cycle of software development and improve the control of modeling complexity. Indeed, MDE is considered to reduce the complexity of existing systems by using a model of high level of abstraction. Also, there is a particular variant of MDE, which is MDA [1] (model-driven architecture), proposed by object management group (OMG) [2]. MDA is a process of software implementation. There are three types of models [2] in MDA: CIM (computation-independent model), PIM (platform-independent model), and PSM (platform-specific model), which corresponds to the software implementation.

In the software development process, the requirements are generally presented by UML diagrams. Nevertheless, this step is considered as a critical step.

The goal of the requirement transformation process is to guarantee the best requirements modeling for a fluent and easy use of information.

The requirement analysis is a former process, always done manually. In the past decade, analysts have achieved semiautomatic and automatic transformations. The computerization of all the tasks was the main goal of analysts. To realize this, they created a procedure comprising many steps to make automatic or semiautomatic requirement transformations into models, mainly UML diagrams.

In this chapter, we present works related to requirement engineering (RE). In addition, we evaluate these works highlighting advantages and disadvantages of each work. The chapter is organized as follows. In Section 5.2, we present the definition of RE and the different existing analysis techniques. In Section 5.3, we present works that transform requirement into models from the first works to the latest works. Section 5.4 discusses artificial intelligence in RE, followed by a comparative study in Section 5.5. In Section 5.6, we finalize with the conclusion.

## 5.2 Requirement Engineering

### 5.2.1 Definition

Before beginning an in-depth study of engineering requirements, we will examine the meaning of "requirements." Indeed, requirements mean archived needs that are the reason for the existence of a system that responds to these needs. So, RE is the process that satisfies the needs of stakeholders by providing a system. The word "stakeholders" means the customer or the user of the end product (system).

To conclude, RE is a process of transformation of user needs. Figure 5.1 summarizes the workflow of this process. An analysis of the requirements is the first step. Its goal is to understand the requirements, and make improvements and refinements to remove ambiguities. Then, a validation step is necessary; its purpose is to ensure that the meaning of the requirements was not changed. Then, the conception stage (design) is followed by the implementation and finally deployment.

The last three stages mark the change of requirements. It is during these stages that requirements are transformed step by step.

This process consists of several phases, such as analysis and design. During the analysis phase, we distinguish three types of requirements: functional requirements, nonfunctional

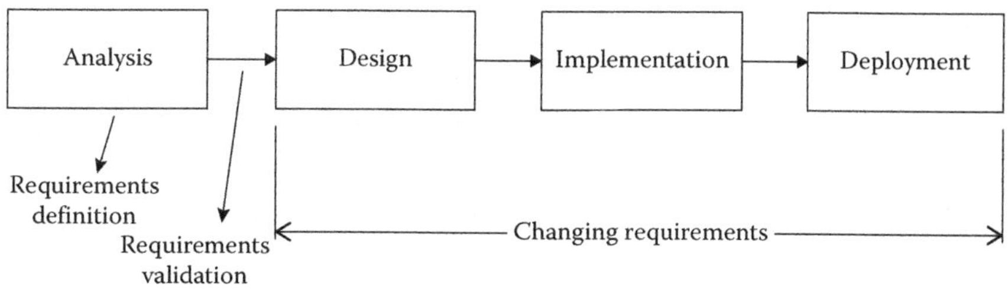


**FIGURE 5.1**
Requirement engineering process.

requirements, and constraints. The design phase is the most delicate and at the same time the most important phase; we try to choose the best line for best results. Assimilating MDE to RE is the right alternative.

IDM appeared to solve design problems and refine the transition from design to execution. In order to ensure a reliable and complete transformation, we go, like the majority of existing approaches, to transform requirements into models and especially into UML diagrams. To achieve this transformation, we can use the already existing analysis techniques for finding information, such as data mining.

### 5.2.2 Requirement Engineering versus Data Mining

The meaning of the term "data mining" is exploration of information. In other words, data mining is the process of data analysis from the data warehouse. The term "data warehouse" refers to the database from which we collect and store information. Data mining is the analysis of existing data, according to a previously specified structure, leading to useful information that is in the form of models or pattern, and then, we can conclude an acquaintance.

Data mining is similar to requirement engineering. Indeed, both processes share the fact that they transform data into models. Except, requirement engineering transforms all data that are written in natural language into models, while data mining selects some data according to a criterion. Then, it transforms the selected data into models. Another point of difference between the two processes is their goal. Data mining has appeared to conclude knowledge from some precise data. Requirement engineering transforms all data requirements to produce a complete system according to the wishes of stakeholders. It deals with all types of requirements: important and not important. Figure 5.2 summarizes all we have said. The upper part of the diagram shows the data mining process. On the other hand, the lower part explains the process of transforming requirements. Data mining extracts data from the data warehouse.

### 5.2.3 UML in Requirement Engineering

This chapter focuses on the transformation of requirements into UML diagrams. As mentioned in Section 5.1, there is a trend toward the transformation of requirements into UML diagrams for reliability, reusability, and so forth.

UML is a modeling language. Its principle is to specify the high-level design solution via UML diagrams enriched by attributes and relations and, then, to generate an executable code. It enables dynamic modeling of the system. The UML profile consists of a total of

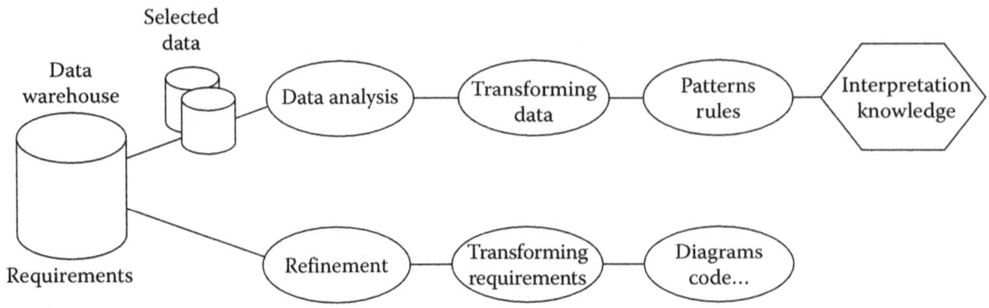


**FIGURE 5.2**
Data-mining versus requirement engineering.

**TABLE 5.1**
UML Diagrams

| Structural Diagrams | Behavioral Diagrams | Interaction Diagrams |
|---|---|---|
| Class diagram | Use case diagram | Sequence diagram |
| Object diagram | State-transition diagram | Communication diagram |
| Component diagram | Diagram of activity | |
| Deployment diagram | | |
| Package diagram | | |

14 diagrams; we quote the 10 most known and used diagrams: class diagram, object diagram, component diagram, deployment diagram, package diagram, use case diagram, state-transition diagram, activity diagram, sequence diagram, and communication diagram. These diagrams are gathered into three categories (structural diagrams, behavioral diagrams, and interaction diagrams) as shown in Table 5.1.

In the literature, several approaches transform requirements into models other than UML diagrams. Other approaches transform requirements into one or two UML diagrams. Although UML diagrams are complementary, each diagram provides specific information. So, we get a complete and clear presentation.

## 5.3 Works That Transform Requirements into Models

Software engineering is a hard and difficult domain to study, especially the engineering requirement phase, which is the most critical to deal with.

To ensure the better transformation of requirements and to obtain a correct and complete output, it is necessary to ensure that the inputs are well presented, since they tend to be ambiguous, understandable, complex, and written by nonexperts.

Considering the importance of refinement of the starting requirements, several studies have focused on this problem to improve requirements.

Over time, approaches and techniques in the domain of transforming requirement have increased—each approach using a different technique. There are approaches based on manual techniques, others on semiautomatic techniques, as well as on automatic techniques. Based on these criteria, we organize approaches that transform requirement engineering into models into three trends: manual techniques, semiautomatic techniques, and automatic techniques.

### 5.3.1 Manual Technique

The first most popular technique is "inspection," [3] named "Fagan inspection," developed by Michael E. Fagan and applied on software development process. It was proposed in 1976. This manual technique is founded on steps followed by the inspection team, as shown in Figure 5.3.

This manual technique needs the intervention of inspectors. So, the skills of inspectors mark the quality of results. For example, if inspectors have to make a decision: acceptation or modification and doing minor modifications or major modifications, this decision, according to their skills, may regard major problems as minor problems. Consequently, they would tend to ignore a major problem with probable defects in the final result. Since this work, every software development approaches used inspections that allow good quality and no defects, as AGILE method.

A few years after the appearance of inspections, Chen introduced the use of linguistic concepts in engineering requirements. In 1983, Chen [4] proposed 11 rules that allow obtaining entity-relationship diagrams.

The organization of templates of English sentences allows getting these rules. The disadvantage of this new approach, which became very popular and was used in many other types of research, is that these rules are not complete, that is, they do not cover all the exceptions and counterexamples that exist in the English language.

As an example, we cite the clause: "In each department, there is an on-line time clock at which employees report their arrival and departure." The analysis proposed by authors for this clause is as follows: this clause present the time, so, it is not important. But, with this analysis and this conclusion, we lose important information that affects the final result.

Among the works that have used inspection, Ackerman's work [5] appeared in 1989. The use of inspection was to identify and delete errors. Authors were satisfied with results using these techniques.

Considering the importance of time, stakeholders will not be satisfied by the performance of these manual techniques that do not guarantee the desired quality and rapidity.

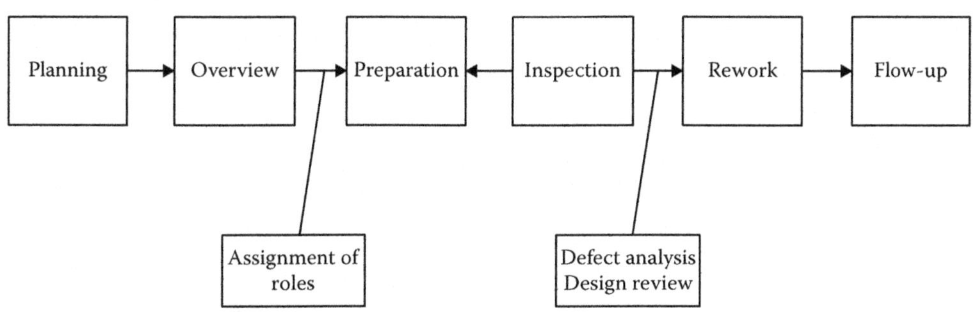


**FIGURE 5.3**
Inspection. (From Fagan, M.E., *IBM Syst. J.*, 15(3), 183, 1976.)

### 5.3.2 Semiautomatic Approaches

To develop and speed up the execution of manual techniques, analysts are following the use of semiautomatic approaches.

In 1996, Börstler [6] and Nanduri and Rugaber [7] presented semiautomatic approach. In fact, in Reference 6 they present an automatic tool to extract a model. The main function of the tool is based on the prespecified key terms in use-case and the two parts of the sentence as verbs and nouns are changed consecutively to behaviors and objects. The approach requests the interaction of the analyst in several phases. Connecting behavior to the object and validating models are finished manually. For example, the performance of the analyzer is noted when it must make a decision as validation of a task. If a mistake is made at this stage, it will be propagated to the end of the process, and so the end result will be wrong.

The other semiautomatic tool is presented by Nanduri [7]. It is a tool that treats syntactic knowledge and needs. The author uses guidelines and parsing rules during the transformation process.

Indeed, an object modeling allows to present guidelines in the form of analysis rules. In this approach, grammatical links are the basic of the parser.

The output of the parser is the input of a postprocessor. This step is proposed by the author [7] who applies the previous guidelines to retrieve objects, attributes, and association. The guideline treatment capacity that this approach can deliver is much reduced, which constitutes the weakness of this method.

The control of natural language is performed by Fuchs [8]. In 1999, Fuchs developed *A*ttempto *C*ontrolled *E*nglish (*ACE*), concerning the control of natural language. Users have to respect some rules to have a clear text.

The domain of RE and especially the domain of transforming requirements into models continues to develop and approaches are becoming more popular.

In 2000, Kroha [9] developed an approach with promising techniques called TESSI. The principle of this approach is the preprocessing of requirement, such as rephrasing to obtain improved requirement. To identify the amount of work, authors applied metrics. So, the goal of this approach is to obtain a semistructured document. The limit of TESSI is that it does not generate models.

The NLP tool CM-Builder [10] peeped out in the same period of PetrCroha's tool. It is an automatic tool (Figure 5.4) that generates from textual specifications an object-oriented

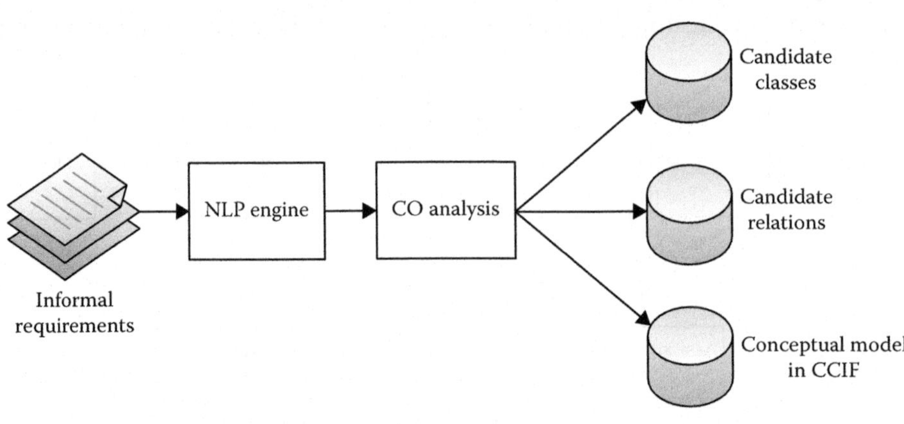


**FIGURE 5.4**
CM-Builder method. (From Harmain, M.H. and Robert, J.G., *ASE*, 45, 2000.)

model. CM-Builder extracts classes from nouns and relationships from verbs. It accepts all types of requirements as input and creates a CDIF file as output. The CDIF file contains object classes, attributes, and their relationships. CM-Builder requests a manual validation from users.

To evaluate this tool, the author has used recall, precision, and specification. The goal of calculating those three metrics is to compare the performance of the actual tool with previous tools. The third metric was defined by the author as follows:

$$\text{Specification} = \frac{N_{\text{extra}}}{N_{\text{key}}}$$

$N_{\text{extra}}$ refers to the volume of extra information that is exact and not found in the answer key $N_{\text{key}}$.

In the paper [10], five uses case were used. The overall scores were 73% recall, 66% precision, and 62% overspecification. These numbers are high compared to numbers of other language-processing technologies.

The author of CM-Builder did not stop at the first version of 2000. In 2003, CM-Builder 2 [11] was developed, which is an improved version of CM-Builder 1. The process is faster and the performance of the new version is more remarkable. The advantage of CM-Builder 2 is that it makes an independent semantic analysis.

In 2001, the reuse of linguistic rules of Chen appeared in the work of Overmyer [12]. The authors have managed to build LIDA, a semiautomatic tool; its principle is summed up in the use of assisted text analysis. The output of this tool is only a UML diagram, which is a class diagram. In the same context of the reuse of rules of Chen, another work appeared—the work of Omar [13].

Another semiautomatic tool based on heuristics appeared in 2004, named ER-Converter. It offers new heuristics and generates ER diagrams (Figure 5.5). In this work, heuristics are collected from all the other works that used heuristics and then gathered in sets, so that every set concerns an ER element.

This tool has 95% as recall and 82% precision. But, it had a limit—it treated only syntactic heuristics.

A new road was followed by Fabbrini [14] when he introduced the *S*oftware *R*equirements *S*pecification (*SRS*) in the transformation of natural language requirements.

The author proposed QuARS (*Qu*ality *A*nalyzer of *R*equirement *S*pecification) to transform requirements into a semiformal model. It is interested in linguistic issues such as ambiguities. In 2001, QuARS was enhanced; the domain of the analysis of model quality improved. The new version [15] detects semantic and syntactic mistakes and runs automatically.

Two works appeared in the same domain and the same period—they are the work of Berry [16] and the work of Rupp [17]. Berry dealt with linguistic issues.

In this section, the approaches studied the outputs, they are interested in the results and verify if they conform to requirements. On the other hand, the previous section was interested in the inputs (starting requirements).

In 2003, the use of feedback in the domain of transforming requirements is present in the work of Briand [18]. In fact, the feedback is an opposite way of the normal direction of the approach. In our context, the feedback allows verifying the synchronization between obtained models and departure requirements. Briand, in his work, used the feedback between two models to verify the changes. In other words, the used feedback can synchronize on model 2 a modification done on model 1.

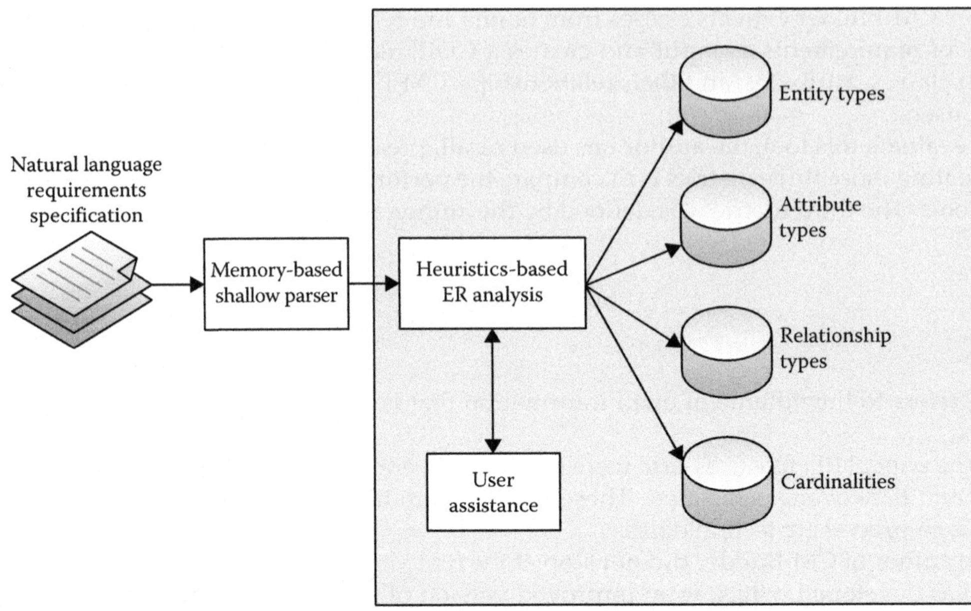


**FIGURE 5.5**
ER-converter method. (From Omar, N. et al., Heuristics-based entity relationship modeling through natural language processing, *Proceedings of the 15th Irish Conference on Artificial Intelligence and Cognitive Science (AICS-04)*, pp. 302–313, 2004.)

Xing [19], in 2005, used the feedback in this context. The proposed approach was UMLDiff. This approach revealed at the end of a treatment a list containing differences between two UML diagrams of the same process.

In the field of processing requirements, several methods have emerged since the beginning, such as inspection, linguistic rules, and NLP (*N*atural *L*anguage *P*rocessing). NLP is a methodology that has appeared since 1989 [20]. It was reused and improved in several types of research such as Hermain's works in the paper [10,21]; also, Dag [22] used NLP to convert the expectations of users into requirements. Kof [23], in 2004, used NLP to extract concepts and then create models without understanding the meaning of requirements.

The use of patterns in requirement engineering was made by Denger [24]. He applied metamodels.

The two papers [25,26], described SENSE, which is a graph-based approach proposed by Tom Gelhausen. The outputs of this approach are UML models.

The author of [27], Deeptimahanti, developed two semiautomatic tools based on natural language requirements (Figure 5.6). The first one, named SUGAR (*S*tatic *U*ML models *G*enerator from *A*nalysis of *R*equirement) [28], generates static UML models. The second tool is named UMGAR (*U*ML *M*odels *G*enerator from *A*nalysis of *R*equirement) [29] and generates static and dynamic UML.

In 2012, Hatem Herchi proposed an approach based on NLP, named DC-Builder [30]. The principle of this approach is to generate only class diagram from requirements (Figure 5.7).

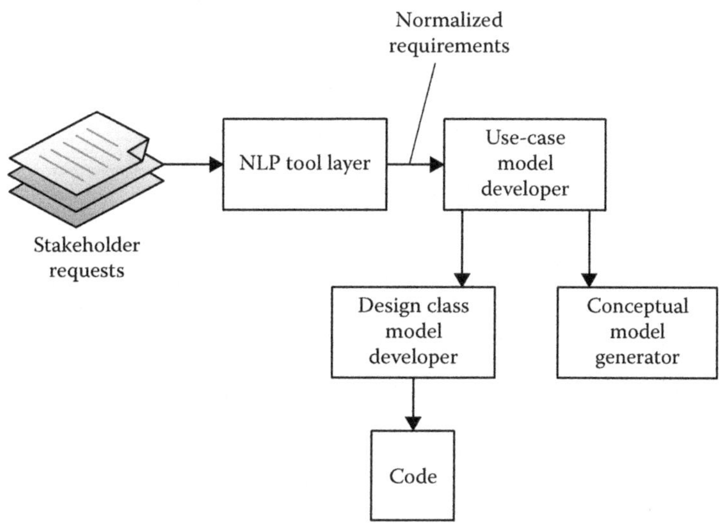


**FIGURE 5.6**
UMGAR tool. (From Deeptimahanti, D.K. and Babar, M.A., An automated tool for generating UML models from natural language requirements, *IEEE/ACM International Conference on ASE*, 2009.)

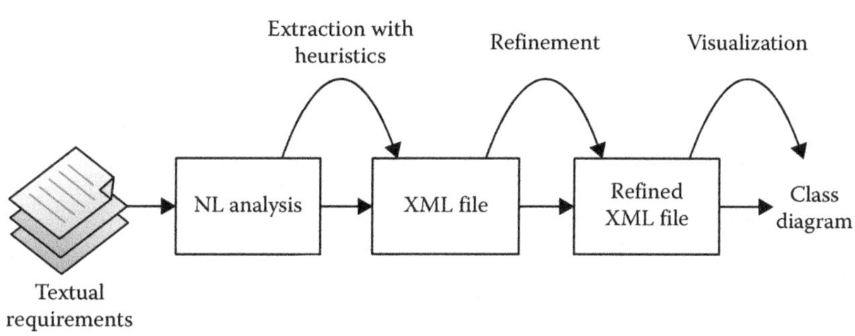


**FIGURE 5.7**
DC_Builder method. (From Herchi, H. and Abdessalem, W.B., From user requirements to UML class diagram, arXiv preprint arXiv:1211.0713.)

### 5.3.3 Automatic Approaches

Generally, the intervention of experts or analysts is a part of the process of transforming requirement into a model, since most of the previous works do not use ontology in their approaches. In fact, the use of ontology in requirement engineering can replace the presence of human intervention. In 2013, another level of approaches was introduced when RECAA (*R*equirements *E*ngineering *C*omplete *A*utomation *A*pproach) [31] was published. It is an automatic tool based on ontology (Figure 5.8).

The newest automatic tool is ABCD [32], proposed in 2016. It is an approach based on NLP and uses patterns and rules to extract class diagram (Figure 5.9).

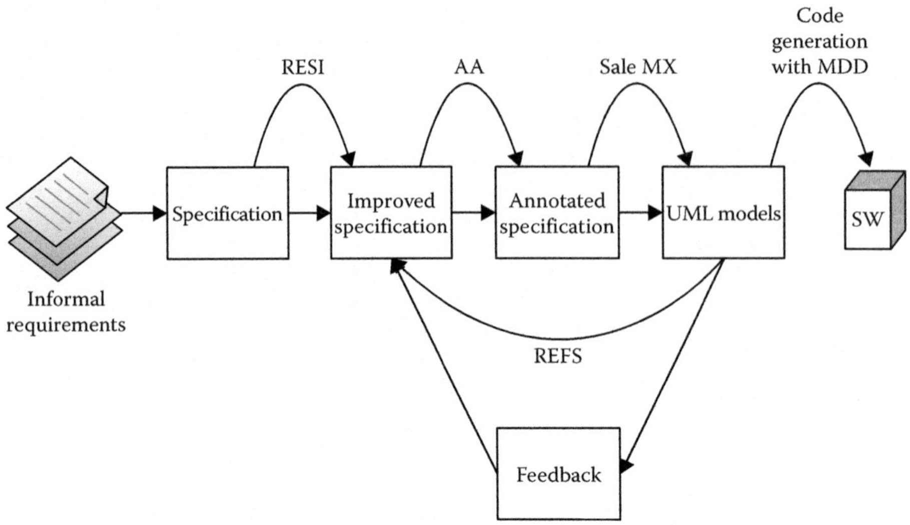


**FIGURE 5.8**
RECAA method. (From Landhauber, M. et al., From requirements to UML models and back how automatic processing of text can support requirements engineering, July 5, 2013.)

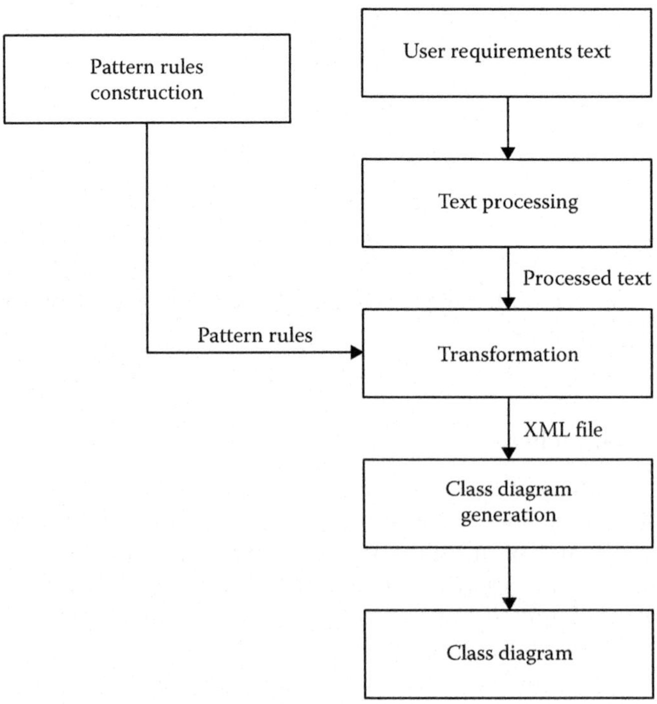


**FIGURE 5.9**
ABCD method. (From Karaa, W.B.A. et al., *Soft. Pract. Exp.*, 46, 1443, 2015.)

## 5.4 Artificial Intelligence in Requirement Engineering

The performance of a machine (computer) depends on the program introduced by the human being. Today, we want machines to react with intelligence, hence the introduction of methods of artificial intelligence.

Developers have achieved the application of AI in RE since the 2000s. In the beginning, it was still arbitrary. Since the 1990s, some attempts, such as [33,34], were concerned with links between these two domains and exposed their overlaps. Also, the application of AI in RE has turned into a necessity to solve ambiguous and incomplete information.

It began with the road map proposed by Nuseibeh and Easterbrook [35]. The authors underlined researches in RE. They exposed, for the first time, the significant developments in RE, and next, they presented their predictions about researches in RE for the subsequent years. The AI was not in their suggestions. We can note that bringing together the two approaches is judged as a new trend. Also, in 2014, a new road map was presented by Sharma [34]. The author exposed the benefits of integrating AI in RE and the use of ontology. He named works that applied AI techniques and identified directions for future researches.

Finally, the last paper [36], which is about the transformation of requirement engineering, shows the importance of AI in this domain. The authors present all the work on ER and show that the use of AI techniques is a necessity to overcome the shortcomings of existing approaches.

## 5.5 Comparative Study

In this chapter, we have mentioned almost all the works that deal with textual requirements, and some works proceed by a refinement of the requirements before applying the transformation process.

Some works applied inspection during the process of transformation; others applied rules, NLP, patterns, graphs, or ontology. And the output of the mentioned approaches is different. There are KML files, models, UML diagrams, and even codes.

The appearance of approaches was not during the same period. In Table 5.2, we sort the used approaches according to their order of appearance.

**TABLE 5.2**

Existing RE Approaches

| Period | Approaches | References |
|---|---|---|
| Beginning of the 1970s | Inspections | [3] |
| Beginning of the 1980s | Linguistic rules | [5] |
| The mid of the year 1995 | • Modeling<br>• Guidelines | [6,7] |
| Beginning of the 2000s | • NLP<br>• Heuristic rules<br>• Patterns | [8–10,21] |
| Years 2007 and 2008 | Graphs | [22,23] |
| Last decade | Ontology | [28] |

From this table, we notice that the use of graphs, patterns, and ontology is a new trend, although many researches use NLP or the rules with very significant results.

The methodologies are developed and improved over time. New techniques appear and old techniques are improved. In fact, we mention the example of the new trend of bringing AI to RE; there is not yet a concrete approach.

Let us talk about the results provided by existing approaches and discuss new research done in order to introduce new concepts.

With time, approaches are improving and results are more satisfactory. However, users are also increasingly demanding.

Starting with limitations of existing approaches, errors made by tools are repetitive. This particular limit reminds us of a concept of artificial intelligence, which is "learning." So, the use of learning concept in the RE approaches makes the tool recognize committed errors so they are not repeated. The goal of the use of AI is to provide to RE tools to react with intelligence—this is the main purpose of many new types of research, such as that of [37].

## 5.6 Conclusion

This chapter detailed the most important approaches and tools in the domain of requirement engineering, especially the transformation of textual requirements. Each named tool or approach has its own process and uses a different methodology. The output is also different from one tool to another. But, the common point between all named approaches is the starting point: textual requirement. The majority of approaches transform textual requirement to models or UML diagrams.

Users are demanding; they require speed, efficiency, and simplicity of use of the tool, which is why we combine approaches in three categories: manual techniques, semiautomatic techniques, and automatic techniques.

The used approaches developed over the time; the use of ontology reduced errors and ambiguities; and the integration of AI in RE may generate good results according to new researches. So, the new trend leads us to smart engineering requirements.

## References

1. Miller, J. and Mukerji, J. MDA Guide, version 1.0.1. Technical report, Object Management Group (OMG), 2003.
2. Bezivin, J. Towards a precise definition of the OMG/MDA framework. *Proceedings of the 16th International Conference on Automated Software Engineering (ASE)*, pp. 273–280. IEEE Computer Society, Washington, DC, 2001.
3. Fagan, M.E. Design and code inspections to reduce errors in program development. *IBM Systems Journal* [Online], 15(3), 183–211, 1976.
4. Chen, P. English sentence structure and entity-relationship diagrams. *Information Sciences*, 29, 127–149, 1983.
5. Ackerman, A.F., Buchwald, L.S., and Lewski, F.H. Software inspections: An effective verification process. *Software, IEEE*, 6(3), 31–36, May 1989.

6. Börstler, J. User-centered requirements engineering in RECORD: An overview. *Proceedings of Nordic Workshop on Programming Environment Research*, pp. 149–156, Aalborg, Denmark, 1996.
7. Nanduri, S. and Rugaber, S. Requirements validation via automated natural language parsing. *Journal of Management Information Systems 1995–1996*, 12(3), 9–19, 1996.
8. Norbert, E., Fuchs, U.S., and Rolf, S. Attempto controlled English. Not just another logic specification language. *Lecture Notes in Computer Science*, 1559, 1–20, 1999.
9. Kroha, P. Preprocessing of requirements specification. In Mohamed, T., Ibrahim, J.K., and Revel, N. (eds.), *Database and Expert Systems Applications, vol. 1873: Lecture Notes in Computer Science*, pp. 675–684, Springer, Berlin, Germany, 2000.
10. Harmain, M.H. and Robert, J.G. CM-Builder: An automated NLbased CASE tool. *ASE*, pp. 45–54, 2000.
11. Harmain, H.M. and Gaizauskas, R. CM-builder: A natural language-based case tool for object-oriented analysis. *Automated Software Engineering*, 10, 157–181, 2003.
12. Overmyer, S., Benoit, L., and Rambow, O. Conceptual modeling through linguistic analysis using LIDA. *Twenty-Third International Conference on Software Engineering*, Toronto, Ontario, Canada, 2001.
13. Omar, N., Hanna, P., and McKevitt, P. Heuristics-based entity relationship modeling through natural language processing. *Proceedings of the 15th Irish Conference on Artificial Intelligence and Cognitive Science (AICS-04)*, GMIT, Castlebar, Irlande, pp. 302–313, 2004.
14. Fabbrini, F. Fusani, M., Gnesi, S., and Lami, G. *Quality Evaluation of Software Requirements Specifications*, 2000, Conference, San Francisco, CA, May 31–June 2 2000, Session 8A2, pp. 1–18.
15. Fabbrini, F., Fusani, M., Gnesi, S., and Lami, G. An automatic quality evaluation for natural language requirements. 2001, *Seventh International Workshop on Requirements Engineering: Foundation for Software Quality*, Interlaken, Switzerland, June 4–5, 2001.
16. Berry, D.M., Kamsties, E., and Krieger, M.M. *From Contract Drafting to Software Specification: Linguistic Sources of Ambiguity: A Handbook*, (Version 1.0) Technical Report. Ontario, Canada: University of Waterloo, Computer science Department, November 2003.
17. Rupp, C. and Sophisten, D. *Requirements—Engineering and Management*, 4th edn. Carl HanserVerlag, Munich, Germany, 2006.
18. Briand, L.C., Labiche, Y., and O'Sullivan, L. Impact analysis and change management of UML models. Technical Report SCE-03-01, Carleton University, Ottawa, Ontario, Canada, February 2003.
19. Xing, Z. and Stroulia, E. Umldiff: An algorithm for object-oriented design differencing. *Proceedings of the 20th IEEE/ACM International Conference on Automated software engineering*, Long Beach, CA, *ASE'05*, pp. 54–65. ACM, New York, 2005.
20. Saeki, M., Horai, H., and Enomoto, H. Software development process from natural language specification. *Eleventh International Conference on Software Engineering*, Pittsburgh, PA, 1989.
21. Harmain, H.M. and Robert, J.G. CM-builder: A natural language-based case tool for object-oriented analysis. *Automated Software Engineering*, 10, 157–181, 2003.
22. Dag, J.N., Gervasi, V., Brinkkemper, S., and Regnell, B. Speeding up requirements management in a product software company: Linking customer wishes to product requirements through linguistic engineering. *Twelfth IEEE International Proceedings of the Requirements Engineering Conference, RE'04*, Kyoto, Japan, pp. 283–294. IEEE Computer Society, Washington, DC, 2004.
23. Kof, L. Natural language processing for requirement engineering: Applicability to large requirements documents. *Requirement Engineering*, 9(1), 40–56, 2004.
24. Denger, C., Berry, D.M., and Kamsties, E. Higher quality requirements specifications through natural language patterns. *Proceedings of the IEEE International Conference on Software-Science, Technology & Engineering (SWSTE'03)*, Herzlia, Israel, p. 80. IEEE Computer Society, Washington, DC, 2003.
25. Gelhausen, T. and Tichy, W.F. Thematic role based generation of UML models from real world requirements. *Proceedings of the ICSC 2007*, Irvine, CA, pp. 282–289, 2007.
26. Gelhausen, T., Derre, B., and Geiss, R. Customizing grgen.net for model transformation. *Proceedings of GRaMoT'08*, pp. 17–24. ACM, 2008, Germany.

27. Deeptimahanti, D.K. and Sanyal, R. An innovative approach for generating static UML models from natural language requirements. In *Advances in Software Engineering*, Communication in Computer and Information Science Springer 30. *ASE '09: Proceedings of the 2009 IEEE/ACM International Conference on Automated Software Engineering. IEEE Computer Society*. Springer, Berlin, Germany, p. 147, 2009.
28. Deva Kumar, D. and Sanyal, R. Static UML model generator from analysis of requirements (SUGAR). *International Conference on Advanced Software Engineering and Its Applications (ASEA)*, Hainan Island, China, 2008, pp. 77–84, 2008.
29. Deeptimahanti, D.K. and Babar, M.A. An automated tool for generating UML models from natural language requirements. *IEEE/ACM International Conference on ASE*, Auckland, New Zealand, 2009.
30. Herchi, H. and Abdessalem, W.B. From user requirements to UML class diagram. arXiv preprint arXiv:1211.0713, 2012.
31. Landhauber, M., Korner, S.J., and Tichy, W.F. From requirements to UML models and back how automatic processing of text can support requirements engineering. Software Quality Journal, Springer US. March 2013, Vol 22, Issue I, pp. 121–149.
32. Karaa, W.B.A., Ben Azzouz, Z., Singh, A., Dey, N., S. Ashour, A., and Ben Ghazala, H. Automatic builder of class diagram (ABCD): An application of UML generation from functional requirements. *Software: Practice and Experience*, 46, 1443–1458, 2015.
33. Meziane, F. and Vadera, S. *Artificial Intelligence in Software Engineering Current Developments and Future Prospects*. IGI Global, Hershey, PA 17033. 10.4018/978-1-60566-758-4.ch014. 2010.
34. Sharma, S. and Pandey, S.K. Integrating AI techniques in requirements phase: A literature review. *IJCA Proceedings on 4th International IT Summit Confluence 2013 - The Next Generation Information Technology Summit Confluence* 2013(2):21–25, January 2014.
35. Nuseibeh, B. and Easterbrook, S. Requirements engineering: A roadmap. *ICSE'00: Proceedings of the Conference on the Future of Software Engineering*, Limerick, Ireland, pp. 35–46. ACM Press, New York, 2000.
36. Abdouli, M., Karaa, W.B.A., and Ghezala, H.B. Survey of works that transform requirements into UML diagrams. *2016 IEEE 14th International Conference on Software Engineering Research, Management and Applications (SERA)*, Towson, MD, pp. 117–123. IEEE, June 2016.
37. Pohl, K., Assenova, P., Doemges, R., Johannesson, P., Maiden, N., Plihon, V., Schmitt, J.-R., and Spanoudakis, G. Applying AI techniques to requirements engineering: The NATURE prototype. *IEEE Workshop on Research Issues in the Intersection between Software Engineering and Artificial Intelligence*, Sorrento, Italy, IEEE Computer, 1994.

# 6

# *Overview of Information Extraction Using Textual Case-Based Reasoning*

**Monia Mannai, Wahiba Ben Abdessalem Karaa, and Henda Ben Ghezela**

**CONTENTS**



**ABSTRACT** This chapter attempts to support the idea of information extraction that can be performed to extract relevant information from texts, using case-based reasoning. In this overview, we summarize several approaches to illustrate this idea. We present also a simple comparison of some systems that use textual case-based reasoning for information extraction.

**KEY WORDS:** *information extraction, medical text, case-based reasoning.*

## 6.1 Introduction

The evolution of the web and the proliferation of cloud computing has presented an unprecedented opportunity for form-free text to coexist online, thus allowing the easy availability of digitalized information. On the other hand, the growing quantity of these data adds increasing complexity from a research, classification, and relevance point of view.

There are many techniques used to alleviate this ambiguity, such as textual case-based reasoning and the different techniques of information extraction. They can help to build a decision support system based on the knowledge extraction process from a set of textual data. For a particular area, it is necessary to identify the knowledge to extract from texts in order to allow decision makers to use them for their analyses of new cases. The system built should help users, firstly, to extract a set of relevant knowledge; secondly, to evaluate whether the extracted knowledge could be applied in the present case, and, finally, to adapt the extracted knowledge to current problems.

Textual case-based reasoning and information extraction are broadly used and applied in several domains and fields such as business, education, research, and medicine. The rapid increase in the amount of readable biomedical text documents has led to an increasing need for textual information identification, relevant data retrieval, data extraction, and knowledge classification. In this context we need the text mining technique as a method that aims to automatically refine the raw information, to display, to establish inherent relationships between similar models and generate appropriate assumptions to user needs.

Taking MEDLINE, the most known biomedical database, as an example, it is obvious that this huge database suffers from a lack of text mining methods and this makes information extraction a more difficult challenge. It is well known that the current amount of digital biomedical data have yet to be fully captured; with this expansion rate, protein–disease-related discoveries are still in their early stages and opportunities ahead for mining the hidden biomedical papers are wide.

MEDLINE database represents a good source to data extraction by experts. However, their efforts often lag behind latest discoveries. Herein, we outline a computational method to uncover possible protein–disease relationships from MEDLINE as a source of data.

This chapter presents and introduces various research works focused on applying textual case-based reasoning and information extraction techniques for the extraction of a pertinent relation concept from medical texts. The chapter includes five sections. Section 6.2 gives an overview of the most important research work related to information extraction techniques from medical texts. Section 6.3 introduces textual case-based reasoning approach. Section 6.4 presents the most important works using textual case-based reasoning for information extraction. Finally, Section 6.5 concludes with some future research directions.

## 6.2 Medical Information Extraction Overview

### 6.2.1 Definition of Information Extraction

Information extraction (IE) is a process of scanning large bodies of text written in natural language, then filtering the information from it. IE could be seen so as the task of natural language processing [1].

The activity of information extraction consists in the retrieval of documents from a collection of text documents available on the web and tagging, afterward, particular terms in the text, to achieve the broad goal of IE method, which is the identification of relevant entities and relationships between them from a wider pool of analyzed text documents [2].

The information extraction task can be defined also as the mechanism of automated creation of a structured representation of extracted information (that can be easily transformed into a knowledge database cord, for example, or any conventional annotation) from unstructured or semistructured machine-readable documents that involves the machine's ability to automatically extract relevant information from unstructured data without the need of manually seeking in a large volume of data to find the exact information required [3].

### 6.2.2 Information Extraction System

The general approach for information extraction calls for the use of programming mechanisms capable of scanning information sources that are considered machine-readable. That is why many systems were developed in order to help to execute searches easily. In the following, we will present some of the IE systems.

#### *6.2.2.1 IE Systems Samples Classified within IE Approach*

One of the most known IE systems is FASTUS (finite state automaton text understanding system), which is a rule-based approach. It is an extracting information system from natural language text that uses nondeterministic finite state mechanisms. Since it was developed back in 1992, it has undergone huge change in coding structure; however, the logic remains identical: a set of cascaded automata that is applied to row data, each pass serving as input to the next one and the end results are combined. Several assessments to this system have shown the reliability and efficiency of FASTUS system in information extraction tasks. This ability allows FASTUS to be more oriented toward information extraction applications and especially information retrieval [4].

In the following we present two different systems for the automatic training approach. The first one is a supervised extraction system, CRYSTAL, which focuses on text handled by a syntactic parser. This system employs a thesaurus and labeled training documents produced by an expert in order to create the extraction rules. Also, it has recourse to inductive learning to obtain the restrictive constraints covering the most similar pair or rules merged together before [5].

For information extraction systems that apply the unsupervised learning, we present herein the AutoSlog-TS as an example, which is an extension of AutoSlog. This system uses a training corpus to provide extraction patterns for the input data using heuristics. Then, it states the reliable patterns by means of statistics and evaluates it, and finally, ranks it according to its statistics relevance [6].

#### *6.2.2.2 IE Systems Classified within Type of Data*

IE systems can be categorized following entry data that are using structured, semistructured data, and unstructured data [7].

*Structured data*: These mainly describe relational databases or data organized in semantic entities. Also, those entities presented in predefined length inside data schema are grouped and have same attributes within one group.

*Semistructured data*: These predominately describe XML, but they may surpass this definition to other forms (tables, databases, file systems, etc.). Here, entities may differ from each other in the same class by their attributes while grouped together, and attributes in the same class can be different and their order not relevant.

*Unstructured data*: These refer to any type of data that do not have a predefined data model or format of the sequence, for examples such raw data may include sound, image, text, and so on. Natural language understanding techniques are required to examine the text and pull out relevant information.

### 6.2.3 Medical Information System

In order to support the discovery of new relationships between concepts in MEDLINE, several efforts are focused on extracting, automatically, associations between concepts. These approaches are often limited to exploring relationships between two concepts, such as drugs–disease associations and disease–gene relations.

In 1991, Sperzel et al. [8] conducted an experiment to investigate the feasibility of using the unified medical language system (UMLS: a repository of biomedical vocabularies) resources to link databases in clinical genetics and molecular biology. References from MIM (Mendelian inheritance in man) were lexically mapped to the equivalent citations in MEDLINE.

In 2004, Gall and Brahmi [9] tested the capabilities of the EndNote search engine, by retrieving citations from MEDLINE and their importation into EndNote. A citation management software package using EndNote version 7.0. Ovid MEDLINE and PubMed were selected for the comparison. Another work in 2005 proposed a technique to rank files from the biggest medical database, MEDLINE, using the data mining technology. The process was founded on a new associative classification technique that treats recurrent topics and most importantly multilabel characteristic of the MEDLINE data [10]. In 2006, Al-Mubaid and Nguyen [11] adapted information-based semantic similarity measures from general English and applied them to the biomedical domain to measure the similarity between biomedical terms.

In 2007, Névéol et al. [12] reported on the latest results of an indexing initiative effort addressing the automatic attachment of subheadings to MeSH main headings recommended by the NLM's Medical Text Indexer. In 2008, a new search concluded whether the information retrieval efficacy is upgraded by making every part of a structured abstract a separate searchable field [13]. In 2009, Humphrey et al. [14] evaluated and compared the systems against a gold original of humanly assigned class for one hundred MEDLINE file, employ six steps selected from trec_eval.

A variety of methods have been published by different authors using different approaches for exploring the relationships between biomedical concepts. Some studies developed systems based on co-occurrence grouping comentioned drugs, diseases, genes, and genomic variations [15]. Some work [16] used the cooccurrence of disease and drug in MEDLINE abstracts to find out drug and to construct a network explaining disease and drug relation. In 2012, Yeganova et al. [17] described and compared two methods for automatically learning meaningful biomedical categories in MEDLINE. Rather than imposing external ontologies on MEDLINE, the methods allowed categories to emerge from the text. In 2013, Jimeno et al. [18] built translation data sets in the biomedical domain that could easily be extended to other languages available in MEDLINE. These sets could successfully be applied to train statistical machine translation models.

The most straightforward approaches detecting medical concept relationships are hybrid approaches. For example, Bchir and Karaa [19] suggested an approach to extract disease–drug relations using machine learning combined with natural language processing: in a first step, they employed natural language processing techniques for abstracts' preprocessing and extracting a set of features from the preprocessed abstracts. In the second step, they extracted a disease–drug relation using machine learning classifier. The Anno-pharma system introduced by [20] used natural language processing techniques, ontology, and dictionaries to detect the substances responsible for adverse reaction on the organs of the human body. In the following, a new methodology has been presented for the extraction of the hidden relationships from MEDLINE. In 2014, Kwon et al. [21] determined the value and efficacy of searching in biomedical databases for systematic reviews. They suggested expanding the range of databases to search in Embase, CINAHL, and so on. A systematic review was conducted on the efficiency of ward closure as an infection control practice. The search strategy was developed by librarians. YK and SEP searched Ovid MEDLINE including in process and other non-indexed Citations, Ovid Embase, CINAHL Plus, and Cochrane Database of systematic reviews (CDSR), LILACS, and IndMED for any study type discussing the implementation of ward closure in the case of an outbreak. In the following table, we present a comparison of some researches related to the relations' extraction from MEDLINE abstracts. Recently, many types of research work related to relations' extraction from MEDLINE abstracts literature have concentrated on specific types of relations; nonetheless, this will restrict the nature of the extracted relations.

## 6.3 Textual Case-Based Reasoning

Artificial intelligence helps us to solve different type of problems in various domains using case-based reasoning because it is a powerful method. The case-based reasoning was defined by many researchers [22,23] as a methodology for solving new problems by adapting a previous solution already used to solve old problems. The ambition of learning is one of the most important targets of case-based reasoning. There are several models for case-based reasoning cycle such as [24–28]. A famous model is proposed by Aamodt, comprising four phases; retrieve similar cases, reuse directly the case solution, revise the proposed solution, and, finally, retain the target case in the case base for future use as shown in Figure 6.1.

Research in a new field make appear textual case-based reasoning as a subfield of case based reasoning (CBR), were the knowledge sources are available in the textual format. It is the case-based reasoning technique that uses the case-based reasoning methodology of solving problems, where experiences are especially picked up from the text. The objective

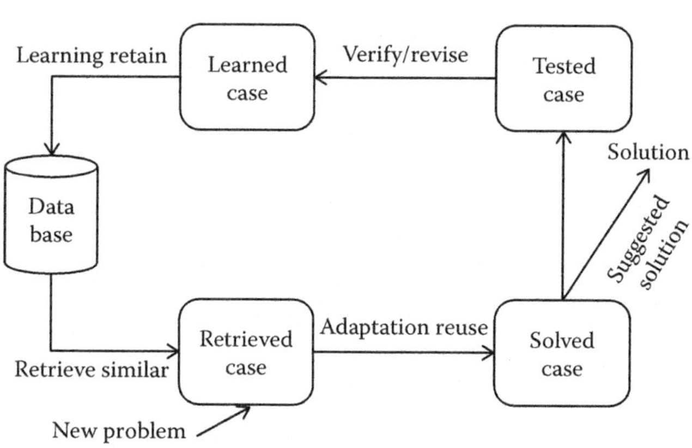


**FIGURE 6.1**
Case-based reasoning cycle.

is to use these textual acquaintance sources in an automated or semiautomated manner for supporting problem-solving via case comparison [29]. The knowledge illustrated in a textual case-based reasoning is very difficult, since it contains complicated grammatical terms and concepts on different topics that often appear in the same case. The target of textual case-based reasoning is to allow classic CBR method to handle reasonably and directly cases interpreted as text. It allows also access to the pertinent textual cases, extract it and affect indices to the textual cases, so that can be collected in the future, or to use the textual cases to reason interpretively about a problem.

**TABLE 6.1**

Textual Case-Based Reasoning System

| Name of System | Role | Reference |
|---|---|---|
| SCALIR | • Developed before the word TCBR was introduced.<br>• Looked to illustrate legal cases texts in a network structure, with favorable results. | Rose [30] |
| SPIRE | • Used a small set collection of excerpts related to its indexing concepts to situate the most hopeful text going on in a new hidden text. | Daniels and Rissland [31] |
| FACIT | • Used natural language processing to derive a profound, logical representation. | Gupta and Aha [32] |
| SMILE+IBP<br>**SM**art Index **LE**arner + Issue-**B**ased **P**rediction | • Developed to leverage an existing collection of cases and corresponding case texts.<br>• Provides response to legal disputes and cases.<br>• IBP is a basic hybrid case/rule-based algorithm that provides the result of legal cases given a textual summary. | Brüninghaus and Ashley [33] |
| ECUE: **E**mail **C**lassification **U**sing **E**xamples | • Anti-spam filter base case,<br>• Based on two variants: the first, the feature-based distance measure, and second, the feature-free distance measure. | Delany and Bridge [34] |
| SOPHIA-TCBR<br>**SOPH**isticated **I**nformation **A**nalysis for **TCBR** | • A textual CBR system that provides a way to combine textual case of a semantic way.<br>• The process of knowledge discovery has five steps: Case knowledge discovery. Narrow theme discovery. Similarity knowledge discovery. Case assignment discovery. Internal cluster structure discovery. | Patterson et al. [35] |
| WikiTAAAble | • The process of knowledge discovery consists of a set of textual recipes described by a title, loosely structured ingredients, and a set of instructions for the preparation list.<br>• The WiKITAAABLE is an extension of the system TAAABLE.<br>• The architecture is formed by WIKITAAABLE semantic wiki that provides easy to implement technical solutions and allows automatic updating of knowledge base through reasoning engine result from the event. | Cordier et al. [36,37] |

**TABLE 6.2**

Advantages and Disadvantages of Textual Case-Based Reasoning System

| Name System | Advantage | Disadvantage |
|---|---|---|
| SMILE+IBP | Permit integrating indexing and text cases and reasoning to identify the principal question. | Problem text indexing case: does not meet the requirements requested by lawyers. |
| ECUE | Replies if spam or not. | The size of case base: if the database size increases it makes difficult search and causes a waste of time. |
| SOPHIA-TCBR | The number of clusters is automatically set; it is discovered by the system. | It has no mechanism for identifying the order of words, which is determined by natural language processing techniques; these techniques are delicate and costly in computing. |
| WikiTAAABLE | • Simple to realize.<br>• The possibilities of adding easily new recipes.<br>• The possibilities of correcting badly commented recipes. | • Redundancy in the ontology.<br>• The free modification of the ontology by any user can raise the risk of introducing incoherence into the knowledge. |

### 6.3.1 Textual Case-Based Reasoning System

Numerous techniques for textual case-based representation are applicable. Rissland and Daniels [38] concentrated on developing a model that helps to map textually expressed cases into the kinds of structured representations used in CBR systems such as SPIRE. In Reference 39, they decompose the text into information entities. In Reference 40, The FAQFinder, a question–answer system was developed by Burke et al., beginning with a classic information retrieval approach founded on the vector space model, where cases are compared as term vectors with weights based on a term's frequency in the case versus in the corpus. Also, the system puts the accent on the similarity between words, which is based on the concept of hierarchy in WordNet. We found also mixed representations between textual and nontextual features in Reference 41. In the following table (Table 6.1), we summarize the most famous textual case-based reasoning systems.

Table 6.2 compares some textual case-based reasoning systems.

## 6.4 Textual Case-Based Reasoning for Information Extraction

A fundamental difference between textual case-based reasoning and information extraction is that information extraction techniques usually are not much concerned with semantic information or heart of domain knowledge regarding problem-solving. On the other hand, textual case-based reasoning processes try to introduce a field of knowledge. Its indexing and collecting techniques use area-precise, question–solution knowledge, as well as more general knowledge, to handle texts to help readers in solving specific problems [42,43]. The specialists of information extraction tend to turn down such field-specific techniques as ad hoc or textual case-based reasoning system. Information extraction methods, which convert a set of documents into more structured statements, have been used in textual case-based reasoning systems. For example, Weber et al. [44] used a variant of IE techniques called template mining to extract information directly from a text when there is an automatically recognizable pattern.

In the following description, we present and introduce the different research works focused on applying textual case-based reasoning and information extraction techniques for extraction of a pertinent relation concept from a medical text. Case-based reasoning is a technique used for building medical systems. Until now, the majority of medical case-based reasoning systems are founded on results of measurements or tests to build cases representations. In order to allow flexible and efficient access to relevant information, analyst experts need text treatment. In this subfield, textual case-based reasoning has been used to facilitate this treatment for medical applications. Many works are presented [45–47] that describe how to use case-based reasoning for information extraction in the medical domain. For example, in Reference 46, the authors used textual case-based reasoning technique to get information from structured documents. In other research work, features vectors are used to build structural cases for recuperating textual cases, where it extracts semantic relationships by the way of association in Reference 48. The similarity of medical cases in MEDLINE has been presented, applying a vector space model and retrieval system using cosine similarity and manual weighting for full-text document [49]. The benefit of mixed textual case-based reasoning and information extraction methodologies is in treating big case-bases in the biomedical domain.

## 6.5 Conclusion

Textual case-based reasoning is a subfield of case-based reasoning, which is an intelligent artificial method. In this chapter, we summarized some of the fundamental concepts of information extraction, and then, textual case-based reasoning was presented, such as retrieval, reuse, and so on. The purpose is to put the textual case-based reasoning and information extraction domain in their contexts. For this reason, we present different research works focused on the methods or techniques for textual case-based reasoning and information extraction. Overall, we can conclude that the methods used to improve the quality of extracting information are rapidly growing.

## References

1. Grishman, R. Information extraction and challenges information extraction a multidisciplinary approach to an emerging information technology. *Lecture Notes in Computer Science*, 1299, 10–27, 1997.
2. Bunescu, R., Monney, R., Ramani, A., and Marcotte, E. Integrating co-occurrence statistics with information extraction for robust retrieval of protein interaction from medline. In *Proceedings of the HLT-NAACL Workshop Linking Natural Language Processing and Biology: Towards Deeper Biological Literature Analysis*, New York, pp. 49–56, 2006.
3. Kauchak, D., Smarr, J., and Elkan, C. Sources of success for information extraction methods. *The Journal of Machine Learning Research*, 5, 499–527, 2004.

4. Appelt, D.E., Hobbs, J.R., Bear, J., Israel, D., and Tyson, M. FAUSTUS: A finite-state processor for information extraction from real-world text. In *Proceedings of IJCAI*, Chambéry, France, 1993.
5. Soderland, S. Fisher, D., Aseltine, J., and Lehnert, W. CRYSTAL: Inducing a conceptual dictionary. In *Proceedings of the 14th International Joint Conference on Artificial Intelligence*, Montreal, Quebec, Canada, pp. 1314–1319, 1995.
6. Riloff, E. An empirical study of automated dictionary construction for information extraction in three domains. *Artificial Intelligence Journal*, 85, 101–134, 1996.
7. Kaiser, K. and Miksch, S. Information extraction: A survey. Technical Report Asgaard-TR-6, Vienna University of Technology/Institute of Software Technology, Wien, Austria, 2005.
8. Tuttle, M.S., Sherertz, D.D., Olson, N.E., Nelson, S.J., Erlbaum, M.S., Sperzel, W.D., Abrabanel, R.M., and Fukker, L.F. Biomedical database inter-connectivity: An experiment linking MIM, GENBANK and meta via medline. In *Annual Symposium on Computer Application [sic] in Medical Care*, pp. 190–193, 1991.
9. Gall, C. and Brahmi, F.A. Retrieval comparison of EndNote to search MEDLINE (Ovid and PubMed) versus searching them directly. *Medical Reference Service Quaterly*, 23, 25–32, 2004.
10. Rak, R., Kurgan, L., and Reformat, M. Multi-label associative classification of medical documents from MEDLINE. In *Proceedings of the Fourth International Conference on Machine Learning and Applications*, Los Angeles, CA, 2005.
11. Al-Mubaid, H. and Nuguyen, H.A. Using medline as standard corpus for measuring semantic similarity in the biomedical domain. In *Proceedings of the Sixth IEEE Symposium on Bioinformatics and Bioengineering*, 2006.
12. Névéol, A., Shooshan, S.E., Mork, J.G., and Aronson, A.R. Fine-grained indexing of the biomedical literature: MeSH subheading attachment for a MEDLINE indexing tool. In *AMIA Annual Symposium Proceedings*, Chicago, IL, pp. 553–557, 2007.
13. Booth, A. and O'Rourke, A. The value of structured abstracts in information retrieval from MEDLINE. *Health Libraries Review*, 14(3), 157–166, 1997.
14. Humphrey, S.M., Névéol, A., Gobeil, J., Ruch, P., Darmoni, S.J., and Browne, A. Comparing a rule-based versus statistical system for automatic categorization of MEDLINE documents according to biomedical specialty. *Journal of American Society of Information Science and Technology*, 60(12), 2530–2539, 2009.
15. Garten, Y. and Altman, R. Pharmspresso: A text mining tool for extraction of pharmacogenomic concepts and relationships from full text. *BMC Bioinformatics*, 10(2), 1–9, 2009.
16. Li, J., Zhu, X., and Chen, J.Y. Building disease-specific drug-protein connectivity maps from molecular interaction networks and PubMed abstracts. *PLoS Computational Biology*, 5(7), e1000450, 2009.
17. Yeganova, L., Kim, W., Comeau, D.C., and Wilbur, W.J. Finding biomedical categories in Medline®. *Journal of Biomedical Semantics*, 3(Suppl 3), S3-S, 2012.
18. Jimeno, Y.A., Prieur-Gaston, E., and Neveol, A. Combining MEDLINE and publisher data to create parallel corpora for the automatic translation of biomedical text. *BMC Bioinformatics*, 14(1), 146, 2013.
19. Bchir, A. and Karaa, W.B.A. Extraction of drug-disease relations from MEDLINE abstracts. In *World Congress on Computer and Information Technology (WCCIT)*, Sousse, Tunisia, June 22–24, 2013.
20. Benzarti, S. and Karaa, W.B.A. AnnoPharma: Detection of substances responsible of ADR by annotating and extracting information from MEDLINE abstracts. In *2013 International Conference on Control, Decision and Information Technologies (CoDIT)*, Hammamet, Tunisia, May 6–8, 2013.
21. Kwon, Y., Powelson, S.E., Wong, H., Ghali, W.A., and Conly, J.M. An assessment of the efficacy of searching in biomedical databases beyond MEDLINE in identifying studies for a systematic review on ward closures as an infection control intervention to control outbreaks. *Systematic Reviews*, 3, 135, 2014.

22. Riesbeck, C.K. and Schank, R.C. *Inside Case-Based Reasoning*. Lawrence Erbaum Associates, Inc., Hillsdale, NJ, 1989.
23. Ashley, K.D. Case-based reasoning and its implications for legal expert systems. *Artificial Intelligence and Law*, 12, 113–208. Kluwer, Dordrecht, the Netherlands, 1992.
24. Kolodner, J. and Leake, D. A tutorial introduction to case-based reasoning. *Case-Based Reasoning: Experiences, Lessons and Future Directions*. AAAI/MIT Press, Menlo Park, CA, pp. 31–65, 1996.
25. Allen, B. Case-based reasoning: Business applications. *Communications of the ACM*, 37(3), 40–42, 1994.
26. Hunt, J. Evolutionary case based design. *Progress in Case-Based Reasoning, Lecture Notes in Computer Science*, Watson, Ian D (ed), vol. 1020. Springer, Berlin, Germany, pp. 17–31, 1995.
27. Aamodt, A. and Plaza, E. CBR: Foundational issues, methodological variations and system approaches. *AI Communications*, 7(1), 39–59, 1994.
28. Kolodner, J. *Case-Based Reasoning Morgan Kaufmann*. Morgan Kaufmann Publishers Inc., San Francisco, CA, 1993.
29. Weber, R., Ashley, K., and Stefanie, B. Textual case-based reasoning. *The Knowledge Engineering Review*, 20(3), 255–260, 2006.
30. Rose, D. *A Symbolic and Connectionist Approach to Legal Information Retrieval*. Lawrence Earlbaum Publishers, Hillsdale, NJ, 1994.
31. Daniels, J. and Rissland, E. Finding legally relevant passages in case opinions. In *Proceedings of Sixth International Conference on Artificial Intelligence and Law*, Melbourne, Australia, 1997.
32. Gupta, K. and Aha, D.W. Towards acquiring case indexing taxonomies from text. In *Proceedings of Sixth International Florida Artificial Intelligence Research Society Conference*, Florida, 2004.
33. Bruninghaus, S. and Ashely, K.D. Reasoning with textual cases. In Munoz-Aliva, H. and Ricci, F. (eds.), *Case-Based Reasoning Research and Development: Proceedings of the Fourth International Conference on Case Based Reasoning (ICCBR-05)*, Chicago, IL., August 2005. Springer Verlag, Heidelberg, Germany, Lecture Notes in artificial intelligence LNAI 3620, pp. 137–151, 2005.
34. Delany, S.J. and Bridge, D.G. Catching the drift: Using feature-free case-based reasoning for spam filtering. *Seventh International Conference on Case-Based Reasoning (ICCBR)*, Belfast, Northern Ireland, 13–16 August. Weber, R. and Richter, M.M. (eds.), *ICCBR, Volume 4626 of Lecture Notes in Computer Science*, Springer, pp. 314–328, 2007.
35. Patterson, D., Rooney, N., Galushka, M., Dobrynin, V., and Smirnova, E. SOPHIA-TCBR: A knowledge discovery framework for textual case-based reasoning. *Knowledge-Based Systems*, 21(5), 404–414, 2008.
36. Cordier, A., Lieber, J., Nauer, E., and Toussaint, Y. Taaable: Système de recherche et de création, par adaptation, de recettes de cuisine. In *EGC*, Strasbourg, p. 479, 2009.
37. Cordier, A., Lieber, J., Molli, P., Nauer, E., Skaf-Molli, H., and Toussaint, Y. WIKITAAABLE: A semantic wiki as a blackboard for a textual case-base reasoning system. In *SemWiki*, 2009.
38. Rissland, E. and Daniels, J. The synergistic application of CBR to IR. *Artificial Intelligence Review*, 10(5–6), 441–475, 1996.
39. Lenz, M. and Burkhard, H. Case retrieval nets: Basic ideas and extensions. *Advances in Artificial Intelligence*. In Görz, G. and Hölldobler, S. (eds), Springer, Berlin, Germany, pp. 227–239, 1996.
40. Burke, R., Hammond, K., Kulyukin, V., Lytinen, S., Tomuro, N., and Schoenberg, S. Question answering from frequently-asked questions files: Experiences with the FAQ Finder system. *AI Magazine*, 18(1), 57–66, 1997.
41. Wilson, D. and Bradshaw, S. CBR textuality. *Expert Update*, 3(1), 28–370, 2000.
42. Lenz, M. *Case Retrieval Nets as a Model for Building Flexible*. Humboldt University of Berlin, Berlin, Germany, 1999.
43. Burke, R. Defining the opportunities for textual CBR. In *Proceedings of AAAI-98 Workshop on Textual Case-Based Reasoning*, 1998.
44. Weber, R., Ashley, K., and Stefanie, B. Textual case-based reasoning. *The Knowledge Engineering Review*, 20(3), 255–260, 2006.

45. Proctor, J.M., Waldstein, I., and Weber, R. Identifying facts for TCBR. In Brüninghaus, S. (ed.), *Sixth International Conference on Case-Based Reasoning, Workshop Proceedings*. Chicago, IL, August 23–26, 2005, pp. 150–159.
46. Weber, R., Ashley K.D., and Brüninghaus, S.B. Textual case-based reasoning. *The Knowledge Engineering Review*, 20(3), 255–260, Cambridge University Press, Cambridge, U.K., 2005.
47. Weber, R., Aha, D., Sandhu, N., and Munoz-Avila H. A textual case-based reasoning framework for knowledge management application. In *Proceedings of Ninth GWCBR*, Germany, pp. 40–50, 2001.
48. Wiratunga, N., Koychev, I., and Massie, S. Feature selection and generalisation for retrieval of textual cases. In Funk, P. and González Calero, P.A. (eds), *Proceedings of the Seventh European Conference on Case-Based Reasoning*, Springer-Verlag, pp. 806–820, 2004.
49. Shin, K. and Sang-Yong, H. Improving information retrieval in MEDLINE by modu-lating MeSH term weights. *Lecture Notes in Computer Science*, 3136, 388–394, Springer, Berlin, Germany, 978-3-540-22564-5, 2004.

# 7

# *Opinion Classification from Blogs*

**Eya Ben Ahmed, Wahiba Ben Abdessalem Karaa, and Ines Chouat**

**CONTENTS**



**ABSTRACT** Many blogs accumulate large quantities of data reflecting the user opinion. Such huge information may be analyzed automatically to discover the user opinion. In this paper, we present a new hybrid approach for blog classification—CARs—using a four-step process. First, we extract our dataset from blogs. Then, we preprocess our corpus using lexicon-based tools and determine the opinion holders. After that, we classify the corpus using our new algorithm Semantic Association Classification (SAC). The generated classes are finally represented using the chart visualization tool. Experiments carried out on real blogs confirm the soundness of our approach.

**KEY WORDS:** *opinion mining, opinion classification, hybrid opinion classification, polarity, blog, semantic association classification.*

## 7.1 Introduction

Different from traditional media such as newspapers, television, and radio, the social media aims to facilitate social interactions across the Internet. Thus, any person can publish freely information using the social media so they simulate the exchange of ideas, the request of advice, or the expression of opinions.

Several types of social media types exist. In this work, we mainly focus on blogs. Indeed, the blog is a regularly updated website or web page, typically run by an individual or small group, which is written in an informal or conversational style [1]. The blogs are a successful tool for opinion expression. The latter may be deeply analyzed. In this context, opinion mining has emerged recently [2]. Such a research area has mainly concentrated on subjectivity analysis, such as opinions, emotions, or feelings [3–5,6,7,8,9,10]. Thereafter, a polarity [10] (i.e., positive, negative, or neutral) may be given to any opinion. These opinion data now cover an obvious strategic and economic impact because their analysis determines the strengths and the weaknesses of any product, estimates the consumer's feedbacks, and thus increases the income.

In this context, we particularly focus on opinion classification from blogs because the blogs cover all human concerns and are a rich source of qualitative data.

Several approaches classify opinion from blogs. Three main pools may be distinguished: (i) lexicon-based approaches: the classification is performed using lexicon-based dictionary [11]; (ii) learning-based approaches: supervised machine-learning algorithms are applied for opinion classification; and (iii) hybrid approaches: mix of both lexicon and learning-based techniques to detect the class of opinion. The flying over the dedicated literature allows us to mention that the preprocessing step is sometimes ignored although its capabilities to improve the dataset quality and the related derived classes. The use of opinion holders is generally limited to adjectives and verbs. Besides, only two classes of opinion are detected. The neutral class is neglected.

To avoid these drawbacks, we introduce our new approach for opinion classification from blogs called CARS, driven by a four-step process. First, we extract automatically our dataset from blogs. Then, we preprocess our corpus using lexicon-based tools and determine the opinion holders. After that, we classify using our new algorithm Semantic Association Classification (SAC). The generated classes are finally represented using the chart tool.

The remainder of the paper is organized as follows. In Section 7.2, we fly over the related work. Section 7.3 sketches our approach. In Section 7.4, we report our experimental results showing the soundness of our proposal. Finally, Section 7.5 concludes our paper and outlines avenues for future work.

## 7.2 Related Work

Under the opinion mining from blogs, three main pools of approaches may be identified: (i) *Lexicon-based approaches*: Glossaries or dictionaries [12] of opinion are designed in order to list the maximum of words that are holders of opinion. Thus, the generated classes may be two (positive vs. negative) or three (positive vs. negative vs. neutral). (ii) *Machine-learning-based approaches:* The words are often considered similar variables.

The semantic aspect is therefore neglected. Supervised learning algorithms are applied such as the support vector machine method and naïve Bayesian network. (iii) *Hybrid approaches*: The corpus is cleaned using linguistic methods and then classified using supervised-machine-learning techniques.

Harb et al. [13] introduced a new approach for opinion detection based on adjectives. First, an automatic extraction of opinion holders' documents is performed. Second, the authors identify the opinion holders' words. Finally, they classify the corpus using the frequency of opinion holders.

Poirier [14] tests and compares the two main pools of opinion classification approaches, namely, lexicon-based and machine-learning-based. Indeed, the latter seem more efficient than the lexicons-based approaches because they are used to classifying the whole documents, while lexicons-based approaches are restricted to the vocabularies involved.

Rushdi Saleh et al. [15] applied the support vector machine algorithm to classify a set of opinions as positive or negative. They apply this method on three different corpus, namely, the corpus used by [16], prepared by the corpus [17], and a newly generated corpus from Amazon.com.

Table 7.1 shows a comparison of opinion classification approaches with respect to several criteria, namely the preprocessing, the classification, and the evaluation. We can see that the preprocessing step is sometimes neglected, although it is mandatory to enhance the quality of classification [13]. Harb et al. [13] detect opinion from adjectives and Poirier et al. use both adjectives and verbs. However, the adverbs are not considered as opinion holders. Among the approaches, some of the works generated only two classes: positive and negative [13,14] and [P11] and neglect neutral class.

Thus, the critical survey of the dedicated literature points out that we can benefit from the lexicon-based approach to clean our corpus and use a classification algorithm for opinion mining detection. However, we may consider several opinion holders' words, namely, verb, adjective, adverb, and so on. In this context, we introduce our new approach for opinion mining.

**TABLE 7.1**

Comparison of Opinion Mining Approaches

| Approach | Method | Preprocessing | Classification | Evaluation | Limits |
|---|---|---|---|---|---|
| Lexicon-based | Harb et al. [13] | • Tree tagger | • Number of adjectives | • Precision | • Few preprocessing<br>• Using only adjectives as opinion holders<br>• Two classes: positive and negative |
| Machine-learning-based | Rushdi Saleh et al. [15] | • n-gram<br>• Tokenization<br>• Stemming | • Support vector machine | • Precision | • No semantic<br>• Two classes: positive and negative |
| Hybrid | Poirier [14] | • Syntax analysis<br>• Stemming<br>• Minisculization of all characters<br>• Punctuation removal | • Support vector machine<br>• Naive Bayesian network | • Precision<br>• Recall<br>• F-score | • Manual preprocessing<br>• Use of verbs and adjectives as opinion holders |

## 7.3 CARS Approach

Starting from blogs as a data source, we propose a four-step process to drive our architecture (*c.f.* Figure 7.1):

1. *Training corpus extraction* aims to automatically extract from the blog comments expressing positive or negative or neutral opinions.
2. *Preprocessing* cleans the corpus and identifies from the training set the opinion-bearing words.
3. *Classification* aims to automatically classify the opinion using our new algorithm in positive, negative, or neutral polarity.
4. *Visualization* the generated classes are represented using the easy-to-use graphical tool, namely the chart.

This process will be detailed in the next sections.

### 7.3.1 Training Set Extraction Step

Our training corpus is extracted from nawet.org, which is a collective blog moderated by Tunisians. It was launched in 2004. To extract the contents of blogs we applied a web-scraping technique called the "WebHarvest."

#### *7.3.1.1 Data Source Selection*

We start by introducing the target site URL in our configuration file as shown in Figure 7.2.

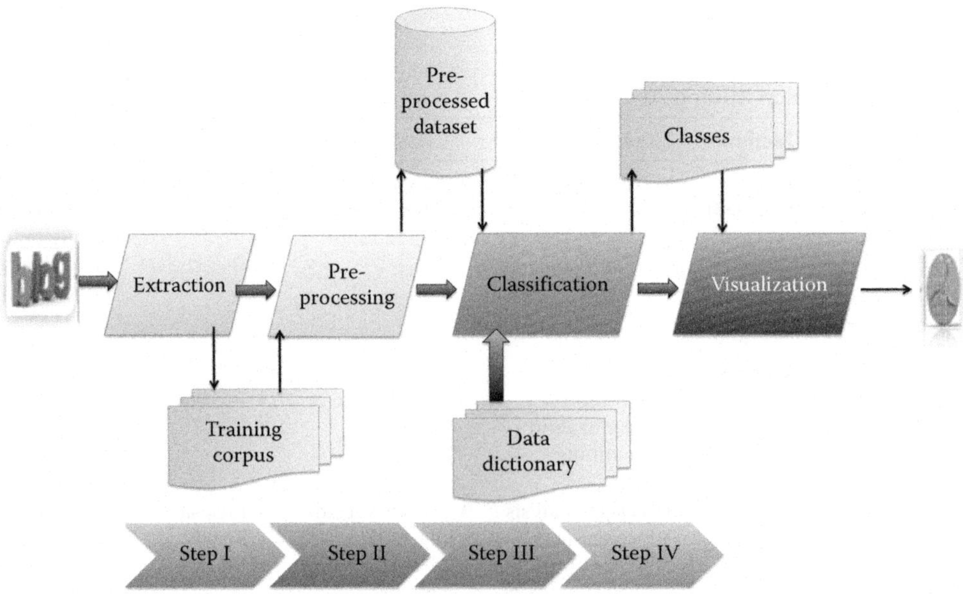


**FIGURE 7.1**
CARS architecture.

```
<var-def  name="getData">
 <html-to-xml>
  <http url="http://nawaat.org/portail/2014/11/27/est-ce-bien-raisonnable-
      de-voter-pour-marzouki/#comments"/>
 </html-to-xml>
</var-def>
```

**FIGURE 7.2**
Pseudo-code of URL input.

The site will be downloaded with the <http> and the downloaded data will be converted from HTML in XML with the <htmltoxml>. Finally, the XML code obtained is stored in a variable that is defined through the element <vardef>.

#### *7.3.1.2 Path Identification*

XPath is a language used to locate a portion of an XML document. Indeed, we use it to extract the required data from the resulting XML document. In our case, we employ the Firefox extension called "Firebug," which determines the path XPath of a website element.

#### *7.3.1.3 Extraction of Final Data*

To extract specific data, we will scan the XML document via <loop>.

#### *7.3.1.4 Load of Extracted Data*

After extracting data from the web, we will load them into CSV file. Figure 7.3 shows the pseudo-code allowing to load comments in a CSV file and Figure 7.4 illustrates an example of generated output.

### 7.3.2 Preprocessing Step

To identify the opinion-bearing words, we perform a preprocessing step. Indeed, three NLP tools [18] are applied: tokenization, stemming, and PosTagging [19].

#### *7.3.2.1 Tokenization*

Tokenization determines tokens, which are the smallest units in the text having meaning. For example, we/are/satisfied/. The output of the tokenization will be the input of the stemming.

```
<file action="append" type="text" path="NAWAT_blog1.csv">
 <template>
  ${blog_title} ¿${blog_author}¿${fdate}¿${fcontent} ${sys.lf}
 </template>
</file>
```

**FIGURE 7.3**
Pseudo-code of comments loading in CSV format.

```
Hédi Sraieb, Moez Joudi & Co. : des économistes ou des zombies?¿TunEconomist¿
Oct27/2014¿Décidément, il y a une épidémie de malhonnêteté intellectuelle en
Tunisie. On ne sait toujours pas grand-chose sur son origine mais il est
clair qu'elle est très contagieuse et elle se transmet, parait-il, par
télépathie : uninvente des idées stupides, une fois ces idées infestent le
cerveau de quelqu'un d'autre, petit à petit elles finissent par manger son
cerveau et il se transforme lui-méme en un zombie mangeur de cerveaux
!J'avais cru porter un coup fatal au il y a quelque mois de ça, mais
apparemment, exactement comme dans la légende, il est très difficile de se
débarrasser de ces créatures. Pire encore, il parait que ce patient vient de
contaminer, qui étaient jusque-là parfaitement saines. En effet, c'est
vraisemblablement Moez Joudi, qui prétend toujours être un économiste et dont
le cerveau semble être infesté par un virus qui lui fait dire en boucle, qui
a persuadé les autres d'écrirepour blâmer la Troïka pour TOUS les maux
économiques que connais la Tunisie en ce moment.Ce qui confirme ma thèse que
ces auteurs sont devenus des zombies obsédés par la Troïka et à la solde des
formations politiques rivales, c'est qu'aucune personne intellectuellement
intègre et mentalement saine ne risquerait sa réputation pour promouvoir une
idée qui ne résistera pas à tout examen rigoureux. En effet, si j'entends
l'épicier du quartier ou mon coiffeur ou n'importe quel autre personne sans
réelle expertise en économie blâmer la Troïka avec toutes ses forces, ça me
fait marrer. Le problème c'est lorsque des personnes comme Sraieb, Joudi &
Co. (mais aussi) font la même chose alors il y a une seule explication : ce
sont des zombies qui promeuvent des idées zombies pour des raisons politiques
et électorales (pour ne pas dire qu'ils sont malhonnêtes !).En économie, la
notion de causalité a occupé les chercheurs pendant très longtemps et c'est
un concept qui n'est pas pris à la légère par les vrais économistes.
Lorsqu'on dit que X cause Y cela veut dire qu'après avoir pris en compte TOUS
les autres facteurs susceptibles d'influencer Y, alors en faisant varier X on
observe systématiquement une variation de Y.Donc, dire que la Troïka est LA
SEULE responsable voudrait dire qu'après avoir pris en compte tous les autres
facteurs (contexte
```

**FIGURE 7.4**
Example of CSV file containing the loaded data.

#### *7.3.2.2 Stemming*

Stemming seeks for the canonical form, called "lemma," for each word. For example, the word "satisfy" exists in several forms such as *satisfied/satisfying*. The canonical form of these words is *satisfy*.

#### *7.3.2.3 POSTagging*

POSTagging is the process of associating to word the grammatical category, namely, adjectives, nouns, adverbs, and so on. We use the TreeTagger [20] grammatical tagger, which provides grammatical labeling texts in several languages.

### 7.3.3 Classification Step

To identify the class of an opinion, we apply the classification rules technique due to its robustness. In this context, we introduce our new algorithm SAC. Inspired from the Apriori [21], it operates in two phases: (i) *frequent itemset generation*, whose objective is to find all the itemsets that satisfy the minimum support threshold; (ii) *class generation*, whose objective is to find the appropriate class according to the data dictionary.

### SAC Algorithm

```
Input: DD: Dictionary data, D: Dataset, s: sentence in D, MinSupp:
minimum support
Output: C: List of derived classes
Begin
 Compute L_1//L1={1-itemset frequent};
 k ← 2
 While L_{k-1} <> Φ
  A_k ←apriori - gen(L_{k-1}) /* New candidates generation */
      While s ∈ D
        A_t = sub-item(A_k, s)
           While a ∈ A_t
        a.count←a.count +1
    L_k ← {a ∈A_k|a.count ≥ MinSupp}
  k ← k + 1
    For each L_i in L_k do
 Check the DD
 Assign for L_i the associated classes c_i
Return ∪_iC_i
End
```

## 7.3.4 Visualization Step

Once the classes are derived, a visualization of the results may be performed as shown in Figure 7.5 using charts.

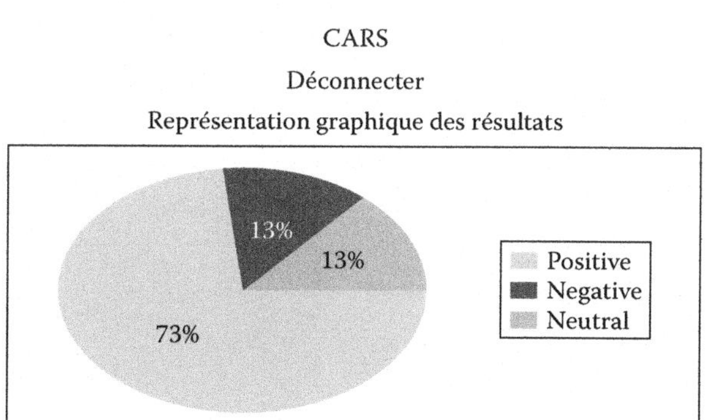


**FIGURE 7.5**
SEQ Figure * ARABIC 5 visualization of detected classes.

## 7.4 Experimental Study

All carried out experiments were conducted on a PC equipped with a 2 GHz Pentium IV and 2 GB of main memory running under Windows XP. Our algorithm is implemented in Java. Figure 7.6 illustrates the pre-preprocessing step. However, Figure 7.7 depicts the data dictionary building.

**FIGURE 7.6**
Preprocessing step of CARS proposal.

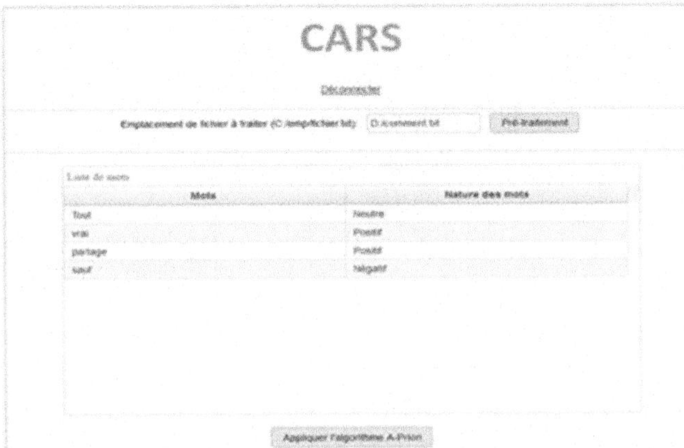


**FIGURE 7.7**
Data dictionary building in CARS proposal.

These experiments have a threefold aim. First, we analyze the scalability of our approach. Second, we put the focus on the performance of our method. Finally, we evaluate the accuracy of CARS.

### 7.4.1 Scalability Analysis

To measure the scalability of CARS, we change the size of the dataset to 40, 80, 100, 150, 250, and 300 Ko. As shown in Figure 7.8, any increase in the dataset size requires more time computation to identify the polarity of extracted opinions.

### 7.4.2 Performance Evaluation

Figure 7.9 plots the runtime of our approach when the minimum support threshold changes. Generally, the runtime of CARS decreases when the minimum support obviously increases.

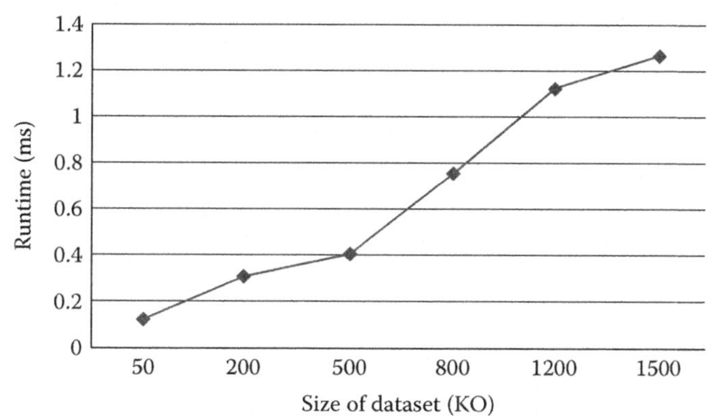


**FIGURE 7.8**
Scalability of CARS.

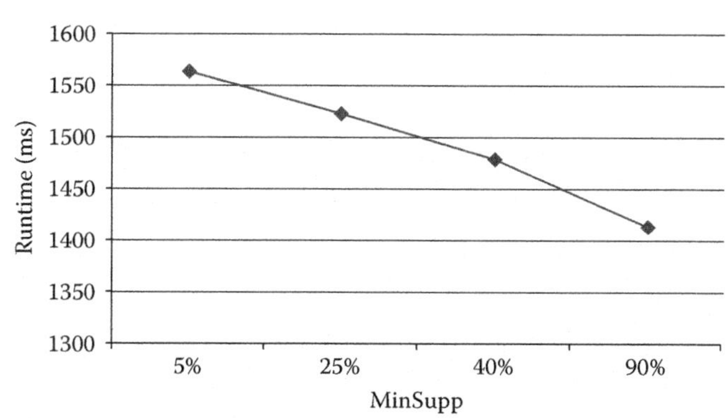


**FIGURE 7.9**
Performance evaluation of CARS with respect to MinSupp.

### 7.4.3 Accuracy Evaluation

To assess the accuracy of CARS, we use two key metrics, namely, the precision and the recall. Indeed, the precision is the number of correctly classified records in respect to all instances in the corpus. Such a metric draws the noise present in derived classes.

$$\text{Precision} = \frac{\text{Number of correctly classified instances}}{\text{Total number of instances}}.$$

However, the recall metric is the number of well-classified instances divided by the total number of correctly classified instances.

$$\text{Recall} = \frac{\text{Number of correctly classified instances}}{\text{Total number of truly classified instances}}$$

Figure 7.10 summarizes the precision rate of our approach in respect to minimum support variation. However, Figure 7.11 shows the recall rate according to the minimum support variation. We notice that the precision and the recall increase with the raise of the minimum support threshold. For example, if the MinSupp is equal to 5%, the accuracy is equal to 0.15 and the recall is equal to 0.19, while the increase of the MinSupp to 90 will provide the maximum of precision reaching 86% and recall reaching 69%.

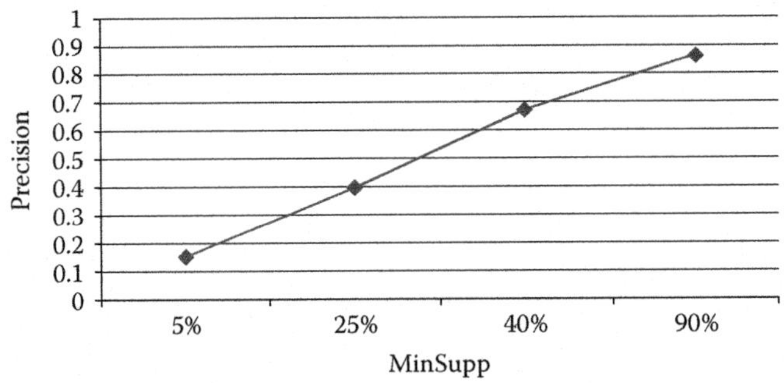


**FIGURE 7.10**
Evaluation of the precision of CARS.

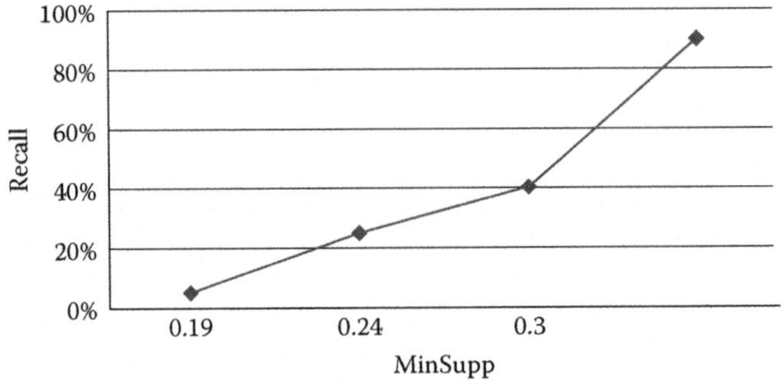


**FIGURE 7.11**
Evaluation of recall of CARS.

## 7.5 Conclusion

In this chapter, we proposed a new hybrid approach called CARS for opinion mining from blogs. Afterward, an automatic extraction of the training corpus, a preprocessing step, is performed to enhance the quality of data and identify the opinion holders. Then, our new method for semantic association classification is applied. The generated classes are finally visualized using ergonomic charts.

Other avenues for future work mainly address the following issues: (i) boosting our classification algorithm by combining with other algorithms such as decision trees and naive Bayesian network to improve the quality of our classification and (ii) considering the ambiguity concern of the language through using ontologies.

## References

1. Bartlett-Bragg, A. 2012. *Blogging to Learn*. University of Technology, Sydney, New South Wales, Australia.
2. Andreevskaia, A. and Bergler, S. 2006. Mining wordnet for fuzzy sentiment: Sentiment tag extraction from wordnet glosses. In *Proceedings of EACL-06, 11th Conference of the European Chapter of the Association for Computational Linguistics*, Trento, Italy.
3. Bayoudh, I. and Bechet, N. 2008. *Blog Classification: Adding Linguistic Knowledge to Improve the K-NN Algorithm*. Université du 7 Novembre à Carthage, Centre Urbain Nord, Tunis, Tunisia.
4. Belbachir, F. 2010. *Expérimentation de fonctions pour la détection d'opinion dans les blogs*. Université de Toulouse, Toulouse, France, pp. 4–6.
5. Cambria, E., Schuller, B., Xia, Y., and Havasi, C. 2013. New avenues in opinion mining and sentiment analysis. *IEEE Intelligent Systems*, 28, 15–21.
6. Feldman, R. 2013. Techniques and applications for sentiment analysis. *Communications of the ACM*, 56, 82–89.
7. Liu, B. 2012. *Sentiment Analysis and Opinion Mining*. Morgan and Claypool Publishers, San Rafael, CA.
8. Ravi, K. and Ravi, R. 2015. A survey on opinion mining and sentiment analysis: Tasks, approaches and applications. *Knowledge-Based Systems*, 89, 14–46.
9. Tsytsarau, M. and Palpanas, T. 2012. Survey on mining subjective data on the web. *Data Mining and Knowledge Discovery*, 24, 478–514.
10. Teresa, M., Martínez-Cámara, E., Perea-Ortega, J., and Ureña-López, L.A. 2013. Sentiment polarity detection in Spanish reviews combining supervised and unsupervised approaches. *Expert System with Applications*, 40(10), 3934–3942.
11. Clark, A., Fox, C., and Lappin, S. 2010. *The Handbook of Computational Linguistics and Natural Language Processing*. Wiley-Blackwell, Maiden, MA.
12. Indurkhya, N. and Damerau, F. 2010. *Handbook of Natural Language Processing*, 2nd edn. CRC Press, Taylor & Francis Group, Boca Raton, FL.
13. Harb, A., Dray, G., Plantié, M., Poncelet, P., Roche, M., and Trousset, F. 2009. Détection d'Opinion: Apprenons les bons Adjectifs!. LIRMM Université Montpellier II, Montpellier, France.
14. Poirier, D. 2011. Des textes communautaires à la recommandation. Ecole Doctorale Sciences et Technologies, Université d'Orléans, pp. 76–79.
15. Rushdi Saleh, M., Martín-Valdivia, M.T., Montejo-Ráez, A., and Ureña-López, L.A. 2011. *Experiments with SVM to Classify Opinions in Different Domains*. SINAI Research Group, Department of Computer Science, University of Jaén, Campus Las Lagunillas, Jaén, Spain.

16. Pang, B. and Lee, L. 2004. A sentimental education: Sentiment analysis using subjectivity summarization based on minimum cuts. In *Proceedings of the ACL*, Barcelona, Spain, pp. 271–278.
17. Taboada, M. and Grieve, J. 2004. Analyzing appraisal automatically. In *Proceedings of the AAAI Spring Symposium on Exploring Attitude and Affect in Text: Theories and Applications*, Stanford University, CA, pp. 158–161.
18. Jurafsky, D. and Martin, J.H. 2009. *Speech and Language Processing: An Introduction to Natural Language Processing, Computational Linguistics, and Speech Recognition*, 2nd edn. Prentice Hall, Upper Saddle River, NJ.
19. Abney, S. 1996. Part-of-speech tagging and partial parsing. In Church, K., Young, S., and Bloothooft, G. (eds.), *Corpus-Based Methods in Language and Speech*. Kluwer Academic Publishers, Dordrecht, the Netherlands.
20. Schmid, H. 1995. Improvements in part-of-speech tagging with an application to german. In *Proceedings of the ACL SIGDAT-Workshop*, Dublin, Ireland.
21. Agrawal, R. and Skirant, R. 1994. Fast algorithms for mining associations rules. In *Proceedings of the 20th International Conference on Very Large Databases*, San Francisco, CA, pp. 478–499.

# Section III

# Multimodal Document Mining

# 8

# *Document Classification Based on Text and Image Features*

**Maram Mahmoud A. Monshi**

**CONTENTS**



**ABSTRACT** In order to increase the effectiveness of multimedia document classification, it is crucial to combine multiple modalities, specifically text and image. Typically, either text content or image content forms the basis for features that are used in document classification. Therefore, researchers are trying to incorporate text and image through multimodal learning and fusion methods. However, there are many challenges involved in this process and thus multimedia document classification has become a research problem of great interest in many domains like the medical field and social media. This chapter provides an extensive survey of recent research efforts on multimedia document classification based on text–image analysis. In particular, the survey focuses on classification background, multimodal learning strategies, multimodal fusion techniques, and multimodal classification applications and challenges. Finally, a conclusion is drawn and some future research directions are recommended.

**KEY WORDS:** *multimedia documents, classifications, textual features, image features, multimodal learning approaches, multimodal fusion approaches.*

## 8.1 Introduction

The number of multimedia documents is increasing rapidly due to the development of social networks, smart phones, digital cameras, and video recorders. Users are creating and sharing documents that contain text, image, audio, or video on a daily basis. This increases the need to classify multimedia documents based on information, otherwise the data in these documents will be useless. Document classification has developed as a sub-branch of the information retrieval (IR) field since the 1990s with various applications such as spam filtering, document indexing, and document ranking [1].

However, most available classification systems utilize one type of information for retrieving and classifying multimedia documents [2]. While text-based image retrieval (TBIR) systems are only text-based such as Google Images, content-based image retrieval (CBIR) systems are only image-based such as the reverse image engine named TinEye [2]. The most popular systems among these retrieval systems are text-based search systems because they retrieve multimedia documents through the document index and metadata such as image names and tags [3].

However, the performance of existing systems such as TBIR and CBIR is limited because they may ignore one or more of the media content. Recent studies have proved that systems that utilize a multimodal approach such as text–image methods provide more accurate results than systems that utilize text-only or image-only features [2,4]. Multimodal information retrieval (MMIR) inevitably combines different retrieval models to search for information on the web in any modality including text, image, audio, and video [5]. For example, Denoyer et al. [1] have proposed a classification method that deals with various content types and considers the structure of electronic multimedia documents. This model applies Bayesian networks to model the document and to integrate different information channels.

This chapter presents an approach for multimedia document classification. This approach takes into account the textual content and image content of these documents. The idea is to represent a document by a set of features to improve classification results.

This chapter is arranged as follows: Section 8.2 explores the state of the art in document classification based on the combination of text features and image features. Section 8.3 evaluates various multimodal learning methods. Section 8.4 examines different multimodal fusion approaches. Section 8.5 presents several applications for the classification based on text–image analysis. Section 8.6 discusses the challenges in the field of multimodal classification and proposes some techniques to overcome these challenges. Finally, Section 8.7 concludes this chapter and suggests future research directions in document classification based on text and image features.

## 8.2 Background

Text modality and image modality have different retrieval models. In order to understand text–image modality, the work done in each modality should be observed thoroughly. The text retrieval problem has attracted increasing attention since the 1940s. For example, MEDLINE system was proposed in 1973 to retrieve online text of medical information [6]. Nowadays, the text retrieval model has two main modes of retrieval, which depend on keywords and categories (vector space modeling), as shown in Figure 8.1 [4,5]. These modes of retrieval have been used in various traditional applications, including content searching and management, text classification, and question answering. These applications are based on statistical analysis or machine learning techniques. In addition, special methods are used to improve the performance of text retrieval systems such as indexing methods to speed up retrieval and query reformulation methods to retrieve relevance information based on user needs [5].

On the other hand, image retrieval methods have attracted increasing attention since the 1990s. For example, the National Aeronautics and Space Administration (NASA), which was explored in the beginning of the 1990s, had the ability to access and manipulate images in different ways [6]. Nowadays, image retrieval has four main modes, which depend on

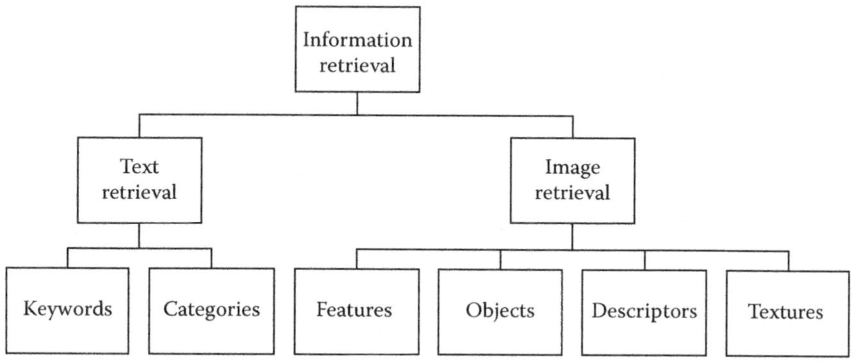


**FIGURE 8.1**
Classic Information Retrieval Methods.

feature, objects, descriptors (retrieval), and pattern recognition (texture), as shown in Figure 8.1 [4,5]. While early systems used image descriptors, in particular color or texture and shape, new systems use bag-of-visual words, scale-invariant feature transform (SIFT), inverted files, and Fisher vectors [5].

However, Multimedia documents combine two or more file formats such as image and text or image and sound [6]. The combination of the file format in a document determines the method of retrieval and classification [6]. Approaches for document classification can be divided into single-modality method, which deals with one information channel such as text only, and multimodal method, which considers more than one channel such as text and image [4]. The earliest efforts in multimodal methods occurred in the 1990s and aimed to analyze texts that surround images. For instance, AltaVista A/V Photo Finder was used to index image collection by applying textual and visual cues [7].

Multimodal classification approaches have been of special interest in recent years as researchers are trying to develop an efficient retrieval system that can satisfy user needs [4,5]. For instance, the classification of the stages of diabetic retinopathy depends on a dataset that consists of images and descriptions written by a medical professional [4]. The classifier may benefit from the two information channels, images and text descriptions, to provide predictions that are more accurate. This example emphasizes the importance of multimodal classifications.

## 8.3 Multimodal Learning Approaches

Each information channel in a multimodal classification has a different representation and correlation structure. While text is represented as discrete sparse word count vectors, an image is represented as outputs of feature extractors or pixel intensities [8]. Therefore, researchers have proposed several approaches to learn from different input channels, including cross-modal canonical correlation analysis (CCA), kernel canonical correlation analysis (KCCA), cross-modal factor analysis (CFA) [9], kernel cross-modal factor analysis (KCFA) [10], semantic matching (SM) [11], semantic correlation matching (SCM) [12], and joint learning of cross-modal classifier and factor analysis method (JCCF) [13]. Table 8.1 compares the most popular multimodal learning methods and points out their strengths and weaknesses.

**TABLE 8.1**

Multimodal Learning Approaches

| Approach | Description | Strengths | Weaknesses |
|---|---|---|---|
| Cross-modal canonical correlation analysis (CCA) | Analyzes mutual information between two multivariate random vectors | Effective in handwriting and face recognition problems | Linear and thus cannot accurately model the relation between two modalities |
| Kernel canonical correlation analysis (KCCA) | Kernelized version of the linear CCA | Outperforms CCA in general problems like fusion of text and image for spectral analysis | Lower classification performance than JCCA |
| Cross-modal factor analysis (CFA) | Projects data of different channels to a shared feature space to classify a text/image directly in this space | Can couple pattern with high variation | Not able to model the relation between two modalities accurately in nonlinearity cases |
| Kernel cross-modal factor analysis (KCFA) | Generalize the linear CFA | Solves nonlinearity issue using kernel trick | Difficult to determine the nonlinear map explicitly |
| Semantic matching (SM) | Supervised method that is based on semantic representation | Gives a higher level of abstraction | Dose not extract maximally correlated features |
| Semantic correlation matching (SCM) | Combines correlation matching (CM) and SM | Enhances the individual performance of CM and SM | Less classification accuracy rate than JCCF |
| Joint learning of cross-modal classifier and factor analysis method (JCCF) | Supervised method that uses factor analysis to project data to a shared space and predict a class label | Improves discriminative ability of both CFA representations and classifiers | Cannot classify documents with multiple images |

CCA, KCCA, and CFA are multimodal subspace methods that are performed in two stages: training and projection. The subspaces are learned in the training stage and then images and texts are projected into these spaces in the projection stage [12]. CFA was proposed by Li, Dimitrova [9] as a suitable tool for many multimedia analysis applications because of its capabilities to remove irrelevant noise. CFA attempts transformation that best represents coupled patterns between different modalities describing the same objects [9]. While CFA is more sensitive to coupled patterns with high variations, CCA is in favor of highly coupled but low variation patterns. However, CFA ignores the supervision information. Therefore, Wang et al. [14] extended CFA by incorporating supervision information to train a class label predictor to use the class label information after projecting the image and text to a shared space by factor analysis.

KCFA uses a kernel trick and extends the linear CFA approach to a nonlinear framework in order to capture the relationship among different variables [10]. The kernel technique is an important tool to design nonlinear feature extraction methods but it is difficult to determine the nonlinear map because the dimensionality of the kernel space is bigger than that of the original input space. Therefore, Wang et al. [10] used a kernel trick to perform nonlinear mapping implicitly in the original input space.

SM is a supervised approach to cross-modal retrieval that depends on semantic representation [11]. SM allows a natural correspondence to be established by mapping images and texts to a representation at a higher level of abstraction. SM also allows the semantic spaces to be isomorphic by representing both texts and images as vectors of posterior probabilities while considering the same set of semantic concepts.

SCM combines SM, which assumes that semantic abstraction is vital for joint text–image modeling, and CM, which assumes that low-level cross-modal correlation is vital for joint text–image modeling [12]. While CM extracts maximally correlated features from text and image, SM constructs semantic spaces through original features to obtain semantic abstraction. Thus, SCM combines CM and SM to improve their individual performance by constructing semantic spaces using the feature representation generated by correlation maximization [12].

JCCF is the first supervised joint cross-modal which incorporates supervision information to project text–image data onto a shared space by factor analysis. Then, this supervised cross-model uses class label information by training a class label predictor in the shared data space [13].

From a practical point of view, the cross-media approach was proposed to jointly model the various topics across visual and textual data in Wikipedia. This proposed method is unique because it imposes an effective optimization technique for feature selection, named L1, to regularize the projection matrix. Thus, only a limited number of related image–text words are associated to each topic [15].

## 8.4 Multimodal Fusion Approaches

Multimedia document classification requires fusion strategies, which combine diverse modality information, to improve classification performance on text and image [2]. There are four types of fusion strategies, namely early fusion, late fusion [5], score fusion [2], and double fusion [2,16]. Table 8.2 compares multimodal fusion approaches and highlights their strength and weakness.

Early fusion is a feature-level strategy that integrates various unimodal features into one representation before performing classification. One simple method to achieve early fusion is to normalize features and then concatenate these features into a unique vector [2].

**TABLE 8.2**

Multimodal Fusion Approaches

| Approach | Description | Strengths | Weaknesses |
|---|---|---|---|
| Early fusion | Fuses features before carrying out classification | Captures the relationship among features more accurately | May overfit the training data and expensive |
| Late fusion | Integrates output of classifiers from various features after classification | Deals with the problem of overfitting the training data | Does not permit classifiers to train on all the data simultaneously |
| Score fusion | Uses score normalization process or pattern classification process | Combines scores from multiple classifiers | Needs a well-formed classification formula |
| Double fusion | Combines early fusion and late fusion | Performs better than early and late fusion when tested on different datasets | Not feasible for use with all features combination when the feature space is large |

Although early fusion captures the relationship among features precisely [16], it may be more complicated and inefficient because it requires extracting features from different data sources and may therefore suffer from data redundancy [2]. Thus, early fusion with the "curse of dimensionality" issue is computationally costly and needs a big set of training data [5]. Principle component analysis, which is a dimensionality reduction approach, can be used to overcome some of the early fusion limitations. In practice, early fusion is used in image applications to integrate texture, color, and shape information such as image classification, image retrieval, image annotation, face recognition, and multimodal biometrics [2].

Late fusion, on the other hand, is a decision-level strategy that processes individual unimodal features and then fuses the outcomes from all systems [5]. Unlike early fusion, late fusion occurs after classification and thus it is easier to perform [16]. Late fusion methods are classified into similarity score approaches, which exploit the similar value between a given query and individual document, as well as rank-based approaches, which exploit the rank of retrieved documents [2]. Average fusion and committee voting are methods used by late fusion to generate a final decision based on the outcomes from multiple classifiers or modalities [16]. Although late fusion is better than early fusion in terms of handling the problem of overfitting the training data, it does not permit classifiers to train on all the data simultaneously.

Score fusion uses multiple modalities to join the scores generated from different classifiers through a rule-based scheme [5]. In order to guarantee that the final decision is based on the significance of each modality and no individual modality will overpower other modalities, score normalization is used to scale the scores generated by multiple modalities in the same range. Another way to conduct score fusion is to consider scores as features in a pattern classification formula [5]. Some researchers like Moulin et al. [2] consider score fusion as a special case of early fusion.

Double fusion was introduced by Lan et al. [16] to incorporate the advantages of early fusion and late fusion. Specifically, double fusion carries out early fusion to obtain a combination of features from subsets of features, and then trains classifiers on each feature combination and performs late fusion on these classifiers' outputs. However, double fusion is computationally not feasible when the feature space is large because the number of all possible future subset combinations is huge, which is $2^{n-1}$, where $n$ is the number of features. Lan et al. [16] attempted to solve this problem by fusing all features in an early step and then conducting late fusion on the results of all single-feature classifiers. Thus, only $n + 1$ classifiers require to be fused.

Choosing the optimal-fusion strategy is an important step in classifying text–image documents more effectively. Researchers concluded that late fusion outperforms early fusion when considering multiple feature types such as text and image because it uses a finely tuned retrieval method suitable to each modality [2,17]. From a deeper perspective, Peng et al. [17] proposed a multimodal fusion that is able of capturing the complementary and correlative relations between images and texts by combining text-only processing and image-only processing to achieve better performance. This multimodal fusion combines the generated results from single-modality methods using ensemble techniques such as liner/maximum rules and logistic regression. Furthermore, Moulin et al. [2] suggested a linear combination model that depends on late fusion and linearly sums up textual and visual scores to obtain the final score as a weighted sum of the generated score from each unimodal system. Researchers proved that using this linear combination model improves results in most multimedia information retrieval systems for combining text with visual features and audio with visual features [2].

## 8.5 Applications

Representing, classifying, and retrieving text–image data have attracted researchers in many professional fields such as machine learning [13], medical domain [4,18], biomedical area [19], and social media [3]. In the area of machine learning, Peairs et al. [20] proposed an automatic method for classifying documents based on texts and images, and then storing these classified documents in the appropriate computer directory. This automatic method applies text analysis and image analysis on the document to build textual and graphical profiles. Then, these profiles are combined using processing logic and a Borda count method to build the document profile [20]. Another automatic machine learning–based system is WebGuard, which classifies web content based on textual, structural, and visual content-based analysis [21]. WebGuard proved that combining textual and structural analysis with visual analysis raises the classification effectiveness to 97.4% accuracy when tested on 400 adult and nonpornographic websites. Hammami et al. [21] claimed that WebGuard's framework could solve other categorization problems of most contemporary websites that combine textual and visual content.

From a medical perspective, medical documents often consist of visual and textual information. Thus, a medical retrieval system needs both text and image information in order to overcome the information limitation in a single source [4,18]. Practically, the multimodality medical image retrieval system (IFM3IRS) application was proposed to utilize text and image information by applying sequential ordering to automatically input the result from text-based processes into visual-based processes [18].

In the biomedical field, combining image and text analyses can improve the effectiveness of document management and classification. This is largely due to the fact that images in the KDD Cup and the TREC Genomics contents provide critical information that cannot be obtained from text and figure captions alone [19]. Therefore, Shatkay et al. [19] have proposed a method to combine image features and text features to specify relevant documents needed by the Mouse Genome Database curators. Their proposed approach, which depends on segmentation, clustering of subfigures, basic text classification, and integration strategy, has supported effective biomedical document classification. Another similar approach for classifying documents in the biomedical domain was suggested by Chen et al. [22], which showed that an image-clustering method can be used to represent a document by visual words. Then, after forming visual words, bag-of-words representation and naive Bayes classifier can be applied to classify a document.

From a social media perspective, text–image co-occurrence data in the web raised the need to classify data based on both text and image features [3]. Social media platforms, such as Facebook, YouTube, and Twitter, process massive amounts of text–image co-occurrence data and thus open up many research opportunities. Practically, an event detection application was proposed to detect events from a Twitter stream using text and image information [23]. The proposed event detection method proved that mining text–image data provides more accurate information with 0.94 accuracy when compared to text-only data with 0.89, or image-only data with 0.86 [23].

Classifying structured documents like XML documents has become an evolving research area that aims to deal with content and structural information as well as different types of content like text and image [24]. Denoyer and Gallinari [24] proposed a method for classifying a structured document that relies on Bayesian networks. Their method can be transformed into a discriminant classifier using the Fisher kernel model and can be extended to handle multimedia content, in particular text and image.

Furthermore, user profiles in social media have led to an explosion in textual metadata like user tags, reviews, comments, and uploader information [7]. The most common metadata is tags because social media services encourage users to annotate multimedia content with tags. These tags have triggered many research topics in multimedia management, in particular tag ranking, tag refinement, and tag information enrichment. In tag ranking, researchers learn tags scores, rank them by neighborhood voting, and then aggregate the results with a score or rank fusion. In tag refinement, researchers eliminate irrelevant content using the ontology in WordNet, refine the tags using visual and semantic similarities, and then expand the tags with their associated synonyms. In tag information enrichment, researchers localize tags into regions, analyze the characteristics of these regions, and then enrich the tag information accordingly [7].

## 8.6 Challenges

Recent studies in the field of text–image retrieval concluded that integrating information from textual features and image features gives superior results than using only one of them [4,6,25]. For example, multimodal classification can be employed on user accounts to classify valid and fraudulent accounts. A fraudulent user may use valid text information but reuse a profile photo, and can thus be recognized as having a fraudulent account through multimodal classification [4].

However, integrating textual and visual similarities raised some issues, including increasing complexity and computing time, manually integrating similarities, and being more expensive [25]. The increasing complexity and computing time in classifying multimedia documents is due to the need to appropriately weight the various modalities to specify ideal fusion strategies [2]. To overcome this challenge, Fisher linear discriminant analysis was applied on text–image documents to learn these weights [2]. This linear weighted fusion solution integrates two or more modalities with less complexity and computing time [2].

Furthermore, most of the available cross-media retrieval systems are not effective over large-scale databases because these systems do not consider scalability issues [7,26]. Therefore, an intermedia hashing model was proposed to overcome the scalability issue by discovering a shared data space for all types of data using intermedia and intramedia consistency and then learning a set of hashing functions for each data type using a linear regression model [26]. The computational overhead involved in applying a culturing algorithm to navigate on a large document collection is another problem associated with documents that include texts and a large number of images [27]. The solution to this problem is to apply a content-based image retrieval (CBIR) system, which concentrates on probability-based topic orientation, semi-supervised document clustering, and interactive text clustering techniques. This CBIR system performs with 94.4% efficiency for multimedia documents because it gets rid of unrelated images to a given query by utilizing major color sets and distribution block signature indexing keys [27].

Another challenging task in multimedia document classification is extracting the text that surrounds an image accurately. Therefore, researchers have been attempting to implement effective systems to extract surrounding text such as the PICITION system, which was developed to access photographs in newspapers, and the WebSeek system, which was established to search images from the web [7].

A multimodal learning model should be effective even in the absence of one information channel. Srivastava and Salakhutdinov [8] have proposed a deep Boltzmann machine (DBM) approach to deal with the missing modalities challenge. The DBM model was tested on documents consisting of texts and images and proved that it was useful for classification and retrieval from both unimodal and multimodal queries. The concept of DBM depends on learning a joint-density model over the space of different input channels and then filling in the absence modalities by sampling from the conditional distribution over them.

## 8.7 Conclusion and Research Directions

This chapter has emphasized the importance of combining multiple modalities, in particular text and image, to raise the effectiveness of document classification. This chapter concentrated on illustrating the relevant background, methods, and current applications and challenges in multimedia document classification. Section 8.2 explored the state of the art in document classification based on both text features and image features. Section 8.3 observed different classification approaches including multimodal learning methods like cross-modal factor analysis (CFA) and Section 8.4 observed multimodal fusion strategies like early fusion. Section 8.5 reviewed classification applications in the medical area, social networks, and other areas that depend on text–image analysis. Section 8.6 discussed associated problems with multimodal classification and suggested some solutions to overcome these problems.

Despite the success of research efforts in the field of multimedia document classification, there are still many open problems that need to be searched and solved. Some proposed future research directions on document classification based on text and image features are developing scalable classification techniques, improving the fronts of multimodal retrieval engines, and classifying documents with multiple images. Developing scalable classification techniques to facilitate large-scale multimedia data is a critical research area because most existing real-world applications require high computation powers and massive amounts of training data [7,26]. Improving the fronts of multimodal retrieval engines can be established by developing new interfaces that can input and output different media data and integrating various media models through new semantic models [5]. Classifying documents with more than one image such as articles can be achieved by representing these images as a single image and then applying the cross-modal classifier method [13].

## References

1. Denoyer, L. et al., Structured multimedia document classification. In *Proceedings of the 2003 ACM Symposium on Document Engineering*, 2003, ACM, Grenoble, France, pp. 153–160.
2. Moulin, C. et al., Fisher linear discriminant analysis for text-image combination in multimedia information retrieval. *Pattern Recognition*, 2014, **47**(1): 260–269.
3. Tian, L., Zheng, D., and Zhu, C., Image classification based on the combination of text features and visual features. *International Journal of Intelligent Systems*, 2013, **28**(3): 242–256.
4. Aryafar, K., *Multimodal Information Retrieval and Classification*. 2015, Drexel University: Philadelphia, PA, p. 131.
5. Bokhari, M.U. and Hasan, F., Multimodal information retrieval: Challenges and future trends. *International Journal of Computer Applications*, 2013, **74**(14): 9–12.

6. Jeong, K.T., *A Common Representation for Multimedia Documents*. 2002, University of North Texas: Denton, TX, p. 113.
7. Zha, Z.-J. et al., Text mining in multimedia. In *Mining Text Data*, Aggarwal, C.C. and Zhai, C. (eds.), 2012, Springer: Boston, MA, pp. 361–384.
8. Srivastava, N. and Salakhutdinov, R.R., Multimodal learning with deep Boltzmann machines. In *Advances in Neural Information Processing Systems*, 2012, pp. 2222–2230.
9. Li, D. et al., Multimedia content processing through cross-modal association. In *Proceedings of the 11th ACM International Conference on Multimedia*, 2003, ACM: Berkeley, CA, pp. 604–611.
10. Wang, Y., Guan, L., and Venetsanopoulos, A.N., Kernel cross-modal factor analysis for multimodal information fusion. In *2011 IEEE International Conference on Acoustics, Speech and Signal Processing (ICASSP)*, 2011, pp. 2384–2387.
11. Rasiwasia, N. et al., A new approach to cross-modal multimedia retrieval. In *Proceedings of the 18th ACM International Conference on Multimedia*, 2010, ACM: Firenze, Italy, pp. 251–260.
12. Pereira, J.C. et al., On the role of correlation and abstraction in cross-modal multimedia retrieval. *IEEE Transactions on Pattern Analysis and Machine Intelligence*, 2014, **36**(3): 521–535.
13. Duan, K., Zhang, H., and Wang, J.J.-Y., Joint learning of cross-modal classifier and factor analysis for multimedia data classification. *Neural Computing and Applications*, 2016, **27**(2): 459–468.
14. Wang, J. et al., Supervised cross-modal factor analysis for multiple modal data classification. In *IEEE International Conference on Systems, Man, and Cybernetics (SMC)*, 2015, pp. 1882–1888.
15. Wang, X. et al., Cross-media topic mining on Wikipedia. In *Proceedings of the 21st ACM International Conference on Multimedia*, 2013, ACM: Barcelona, Spain, pp. 689–692.
16. Lan, Z.-z. et al., Multimedia classification and event detection using double fusion. *Multimedia Tools & Applications*, 2014, **71**(1): 333–347.
17. Peng, Y. et al., Multimodal ensemble fusion for disambiguation and retrieval. *IEEE Multimedia*, 2016, **23**(2): 42–52.
18. Madzin, H., Zainuddin, R., and Sharef, N., IFM3IRS: Information fusion retrieval system with knowledge-assisted text and visual features based on medical conceptual model. *Multimedia Tools & Applications*, 2015, **74**(11): 3651–3674.
19. Shatkay, H., Chen, N., and Blostein, D., Integrating image data into biomedical text categorization. *Bioinformatics*, 2006, **22**(14): e446–e453.
20. Peairs, M., Hull, J.J., and Cullen, J.F., Automatic document classification using text and images. 2006, Google Patents.
21. Hammami, M., Chahir, Y., and Chen, L., WebGuard: A Web filtering engine combining textual, structural, and visual content-based analysis. *IEEE Transactions on Knowledge and Data Engineering*, 2006, **18**(2): 272–284.
22. Chen, N., Shatkay, H., and Blostein, D., Exploring a new space of features for document classification: Figure clustering. In *Proceedings of the 2006 Conference of the Center for Advanced Studies on Collaborative Research*, 2006, IBM Corporation: Toronto, Ontario, Canada, p. 35.
23. Samar, M.A., Suhuai, L., and Brian, R., Fusing text and image for event detection in Twitter. *International Journal of Multimedia & Its Applications*, 2015, **7**(1): 27–35.
24. Denoyer, L. and Gallinari, P., Bayesian network model for semi-structured document classification. *Information Processing & Management*, 2004, **40**(5): 807–827.
25. Buffoni, D., Tollari, S., and Gallinari, P., A Learning to Rank framework applied to text-image retrieval. *Multimedia Tools & Applications*, 2012, **60**(1): 161–180.
26. Song, J. et al., Inter-media hashing for large-scale retrieval from heterogeneous data sources. In *Proceedings of the 2013 ACM SIGMOD International Conference on Management of Data*, 2013, ACM: New York, pp. 785–796.
27. Karthikeyan, M. and Aruna, P., Probability based document clustering and image clustering using content-based image retrieval. *Applied Soft Computing*, 2013, **13**(2): 959–966.

# 9

# *Content-Based Image Retrieval Techniques*

**Sayan Chakraborty, Prasenjit Kumar Patra, Nilanjan Dey, and Amira S. Ashour**

**CONTENTS**

**ABSTRACT** Technology enables the attainment, storage, transmission, and manipulation of a huge compendium of images. Content-based image retrieval (CBIR), interchangeably known as content-based visual information retrieval (CBVIR) and query by image content (QBIC), is a computer application using visual techniques of representation, organization, and search. In large databases, images are systematized by their content without human infringement instead of using annotation. In this case, image retrieval does not settle for keywords or annotations, but is merely founded on feature extraction from the images. The retrieval depends on the precise extraction of characteristics to describe the hidden contents of the images. CBIR retrieves, locates, and displays visually similar images to a specified query from an image database by a set of features and image descriptors. Furthermore, proper querying, indexing, matching, and searching methods are compulsory. To advocate this CBIR technique, pattern recognition, statistical techniques, signal processing, and computer vision are corporately set out.

**KEY WORDS:** *content-based image retrieval, similarity measurement, feature extraction, textures, histogram analysis.*

## 9.1 Introduction

Recent trends in image processing show that content-based image retrieval (CBIR) is one of the most prevalent domains of research. This method can be used in image databases. For example, given the particular texture or color of a target image, the image can be retrieved from the image database. It should be noted that the main objective of using the CBIR [1] technique is to minimize texture descriptors. As previously noted, feature extraction is the crucial step in CBIR. It describes the image with the least required number of descriptors. The most commonly used visual features involved in CBIR [2] are color, shape, and texture. Although texture does not provide the framework required for visual information–based feature extraction, it helps to obtain images based on their texture or pattern. In earlier works, texture information has been investigated using pattern recognition. Wavelet transformation is another method that helps to identify textures inside an image. Histogram analysis is a widely used method to obtain the color component of an image. Conventional color histogram of any image refers to the frequency of each color inside a particular image. Overall, it can be said that CBIR [3] can be used to retrieve images based on their similarities with the help of their features such as shapes, textures, and colors.

CBIR can be defined as the solution to the problem of image extraction [4,5] from large databases, which is encountered in computer vision techniques. Identifying the actual content is the chief goal of CBIR instead of obtaining metadata (e.g., tags, descriptions, or keywords) from an image database that are associated with images. Generally, users are interested only in a small portion of the image database. The main problem arises when the images in the database are unlabeled. This problem [6] is mostly solved using classification along with CBIR. Classification categorizes images from an image database into two classes such as a positive class and a negative class. Often, positive examples are clustered in a certain way, but in the case of negative examples they do not usually cluster as they may belong to any class.

Recent CBIR techniques [7] have mainly focused on features of images in order to address the query during an image database study. These features can again be categorized into two types: low-level and high-level features. Apart from feature extraction, similarity comparison is another major component of CBIR. Similarity comparison helps to retrieve target images from the database. Apart from feature extraction and similarity comparison, clustering techniques also play a major role in CBIR. Clustering algorithms help to organize multidimensional data from large image databases. Usually clustering algorithms use local search techniques to efficiently execute the CBIR process.

Section 9.2 discusses related works of CBIR. The CBIR frameworks are described in Section 9.3. Section 9.4 provides the concept of different CBIR techniques whereas Section 9.5 presents a comparative study of CBIR techniques. The conclusion is reported in Section 9.6.

## 9.2 Related Works

Previously, widespread research was conducted in the CBIR domain. A few studies tried to optimize the framework, whereas others tried to explore various features during the feature extraction procedure. All of these methods are discussed in this section.

In 2006, Das et al. introduced reweighting of features during CBIR. The authors discussed the CBIR framework issues [8] such as large dimensions in feature space and obtaining system parameters from feedback samples during the learning process. In order to minimize these issues, the authors executed various weight update schemes and presented a comparative study. A year later, Katare et al. proposed a novel CBIR system [1] using combined features in multi-object images. In this work, the authors tried out shape segmentation to cope with multiple object scenarios. In order to successfully execute the whole process, the glottovibrogram active contour is used. They proposed a system that automatically initialized the active contours. In 2008, Guldogan and Gabbouj built an algorithm [6] for selecting the features during CBIR. The suggested system aimed at maximizing the semantic image retrieval [9] results. It also decreased the complexity of the retrieval process and improved the multimedia search system stability and usability for these engines' end users.

Vassilieva surveyed the existing content-based [10] image retrieval methods. The paper was a representation of a survey of the most commonly used feature extraction techniques and similarity measurements techniques. Color, shape, texture features, and various techniques related to them were also discussed in this paper. Classification techniques used till 2009 were also presented as a literature survey. In 2010, Zhang and Zou used color and edge detection [2] features during CBIR. In this study, the authors used color index codes for image retrieval with edge detection features as the color attribute's weight of the image. This edge detection feature's values were present inside the color feature's subblock. Later in 2011, Schaefer proposed basic techniques [5] of CBIR. Mainly, basic features of images were used during feature extraction in the CBIR system. Various types of distance metrics were used during CBIR [7] in a novel work done by Patil and Talbar. The work presented a comparative study of six different distance metrics based on the CBIR [11,12] system. They were Euclidean, Canberra, Manhattan, square chord, Bray–Curtis, and square chi-squared. Energy levels were calculated using pyramid-structured wavelet decomposition.

Singh et al. [13] suggested a new technique of CBIR for the dynamic environment problem. In real-time computer vision systems, it is not possible to analyze each image during the storing operation. The authors proposed a system that selected the most useful features in order to analyze the newly stored or received images to reduce the analyzing problem in real time. The system was not only stabilized but also made accurate to analyze images in real time. The algorithm was designed in such a way that the feature vectors were used following the segmentation operation for similarity comparison between the image database and the query. The whole framework was later trained to work for various images present in the database.

Visual features were used to design a CBIR technique proposed by Chang et al. in 2013 that was later optimized [12] employing particle swarm optimization. In 2014, color edge detection and discrete wavelet transformation (DWT) were both used for CBIR [3] by Agarwal et al. A novel technique that combined CBIR, DWT, and color edge detection was proposed in this work, which claimed to be different from the existing methods based on histogram analysis. Recently, Ghuge and Patil proposed an approach [4] based on radeon projection to retrieve images in CBIR. In this work, the authors proposed a CBIR technique that used radon transform and histogram. Radon transform is based on the image intensity's projection along a radial line at a precise angle. Yasmin et al. [11] introduced EI classification in the CBIR technique, which was based on color feature extraction. In this work, images were converted into a minimum of 16 squares up to 24 squares of equal size. Edge detection was later applied to the converted parts followed by pixel classification. Pixel classification in this approach relied on pixels found inside and at the edge of the image.

This section mainly presented a literature survey of the most important techniques that were conducted in the domain of CBIR. There has been a lot of work done previously which has been summarized in this chapter in order to discuss the novel approaches over the years. The earlier discussion clearly denotes that various CBIR techniques exist as well as several components in CBIR frameworks. These are discussed in the following sections.

## 9.3 CBIR Components

Feature extraction and similarity measurement are two main components or rather two main frameworks that are used. Feature extraction plays an imperative role in retrieving images from the database; alternatively, similarity measurement perfectly extracts the target image from a large database. These two components or frameworks are discussed in this section (Figure 9.1).

### 9.3.1 Feature Extraction

Any CBIR framework's base is visual feature extraction. The features present in images can be of two types: visual features and text-based features. Texture features can be further categorized as keywords and annotations. Meanwhile, visual features include colors, shapes, and textures present inside an image. As previously noted, features can be either low level or high level. The key of a CBIR system [14] is the selection process of the features, which mainly represents the image involved. The main reasons behind

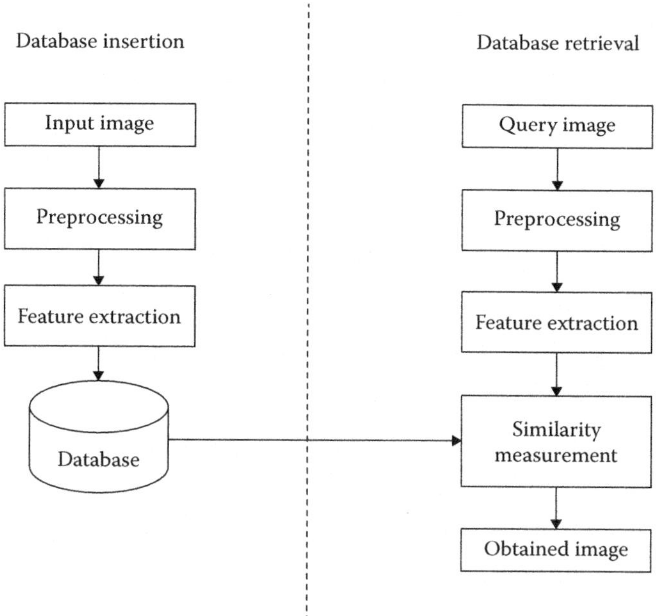


**FIGURE 9.1**
Block diagram of CBIR components and their relation.

multiple approaches toward visual features are the complex composition of visual data and perception subjectivity.

Numeric or alphanumeric representation of attributes of digital images helps to compute the compact representation of the visual features of that particular image. Automated computation of such operations is a key part of feature extraction. Such information is further used during the feature extraction process. Mainly the whole process leads to reduction in dimensionality. An attribute or a feature of any image is not only associated with the visual characteristic but also related to symbolic, semantic, or spatial characteristics. It is possible for each feature to be related with a single attribute or their relation can be with a composite representation of different features or attributes. Features can also be classified in terms of purpose. They can be of general purpose or domain dependent. General-purpose attributes can be employed as a common technique for many feature extraction approaches, whereas domain-dependent features are mainly built for particular computer vision applications. It has been seen in the past that researchers have chosen feature extraction frameworks wisely, as per their proposed system's requirement.

Image representation [15] also plays a major role during feature extraction as the most useful features are carefully chosen to represent a particular image's contents. The approaches to feature extraction are also handled wisely to effectively extract the attributes of the images. Often the feature extraction [16] from an image database is done off-line, which leads to complexity, and therefore becomes irrelevant to the framework.

From the discussion it is quite clear that the sole purpose of CBIR [17] is to obtain images from large databases. Image retrieval can be done with the help of automatically extracted attributes and features by using queries. A query basically explains the complete image or image's parts. Later, according to its similarity with the query, the image is searched and obtained properly. Similarity is also founded on either the total image or parts of that image. Various levels of image descriptions exist in CBIR [18] systems, such as color, texture, and shape. The features that are visual contents of the image are identified as content-dependent metadata. The data related to content semantics are identified as content-descriptive metadata.

During CBIR, the following steps are mainly used:

1. Choosing the target image
2. Executing feature extraction from the target image
3. Obtaining images with similar features that are extracted from the target image

In most of the methods, it has been observed that prior to color feature extraction, the red, green, blue (RGB) image is transformed into a hue, saturation, and value (HSV) color image. The color models are discussed in Section 9.4. This operation is mainly done because RGB doesn't have proper human perception, but HSV does. Afterward, color histograms are accordingly generated for each of the colors. Color histograms are further normalized. Vector feature representation includes these color histogram's values. This process helps to solve the problem created by the combination of similarity measurement.

Color, shape, and texture are the most regularly used features for CBIR feature [19] extraction. These features are categorized into low-level features. Robustness, implementation simplicity, and effectiveness are the main reasons that color is mostly chosen among the available low-level features. As previously discussed, the RGB color model is mainly neglected because of human perception and the images are converted into an HSV or Commission Internationale de l'Eclairage (CIE) color model. Texture is another popular feature that is widely used by researchers. The main intention of using texture as a feature in CBIR is to retrieve [20] the pattern and granularity of surfaces inside the image. Previous works have shown that MPEG-7 standard had a defined set of texture and color descriptors as well as spatial, texture, and histogram-based descriptors to interpret natural images.

In many CBIR frameworks, it has been observed that reducing the semantic gap between human semantics and visual features was considered to be the main target. Object ontology was used to obtain high-level semantic features for CBIR. On the contrary, to obtain low-level features supervised and unsupervised learning has been used.

### 9.3.2 Similarity Measurement

CBIR's main purpose is to enable users to obtain images from a database. The image is denoted by a set of low-level descriptors that is used to measure the distance functions or similarity according to the query. In the past, high-level semantics have been obtained from low-level descriptors. The key to building a stable CBIR framework [21] is to choose the best method to combine the techniques discussed earlier. During similarity measurement, a dissimilarity function is required to measure the similarity index. For every similarity and dissimilarity function, a probability distribution is required. According to statistical independence, these probability distribution–based functions measure the similarity from the given image database. The advancement of technologies has connected the whole world using digital data, which is fast increasing in volume. Hence, multimedia systems are urgently required that can search, access, and explore large amounts of data in real time.

In a multimedia retrieval system, similarity search plays a vital role. This technique is widely used in scientific as well as commercial applications such as near-duplicate detection of images and videos or CBIR-based audio, video, or image retrieval. The main challenges of data objects' inherent properties are collected using feature representation. It has been observed that any similarity measurement [22] framework can define the similarity between the query and the target object from the database. This is done by measuring the distance between the corresponding feature representations. These distance values can be further processed to retrieve mostly similar objects from the database.

## 9.4 CBIR Techniques

The term "content-based" indicates that the search is about the image contents instead of metadata, which are associated with tags, keywords, or descriptions. The word "content" in this context eludes properties like shapes, colors, textures, or any of the image's acquired information. CBIR is necessary because its searches rely on automated image retrieval features. There are a number of image retrieval techniques that are adopted to fulfill this requirement. Some of them are highlighted in the following subsections.

### 9.4.1 Color-Based Image Retrieval

To a great extent, color is a robust background complication along with image size and orientation, which are independent visual features of CBIR. A classic CBIR solution necessitates the construction of an image descriptor that considers (i) an extraction algorithm for encoding image features to feature vectors; and (ii) a matching function for the similarity measurement technique to provide a similarity degree for a given pair of images. The color description techniques are classified into two groups based on yes or no information encoding related to the color spatial distribution, as will be highlighted later.

Descriptors without spatial color distribution include color transformation (color space quantization), color histogram, geometric moments, and color moments.

#### *9.4.1.1 Color Transformation (Color Space Quantization)*

Typically, the image color is characterized through some color model. A color model [23] is delimited in terms of a subspace within that system and a 3D coordinate system, where each color is characterized by a single point. The color space models can be distinguished as user oriented and hardware oriented. The hardware-oriented color spaces, comprising RGB, CMY, and YIQ (luminance and chrominance), are based on the three-color stimulus theory. The user-oriented color spaces, which include HCV, HLS, HSV, CIELAB, and CIE-LUV, are based on the three human precepts of colors, that is, saturation, hue, and brightness. In color space quantization, the color space is reduced from all probable colors to a discrete colors set. Actually, this process of quantization is the same as the reducing color process.

#### *9.4.1.2 Color Histogram*

Descriptors are generally used in image retrieval. The color histogram extraction algorithm can be split into three stages, namely, color space partition into cells, each cell's association to a histogram bin [24], image pixel number counting of each cell, and storing this count in the analogous histogram bin.

#### *9.4.1.3 Color Moments*

To increase the quantization effects of the color histogram, color moments are employed as feature vectors for image retrieval. Thus, color distribution can be characterized by its moments. The majority of the information can be extracted at low moments. Usually, mean, variance, and skewness are used to form the feature vector. Thus, the degree of asymmetry in the distribution can be measured by its skewness.

#### *9.4.1.4 Geometric Moments*

An image moment is the image pixel intensity's definite weighted average that is generally selected to retrieve some significant characteristics. Image moments are convenient to define objects following segmentation.

#### *9.4.1.5 Color Coherence Vector*

The color coherence vectors (CCVs) are formed by calculating the total number of coherent and incoherent pixels for each color. This method is similar to the color histogram method. CCV uses some spatial features and has been proved to be more effective than the classical histogram method.

#### *9.4.1.6 Border/Interior Pixel Classification*

In this method, each image pixel is categorized as an interior/border pixel.

#### *9.4.1.7 Color Correlogram*

The color correlogram is represented as a table indexed by color pairs, where the $m$th entry for $(a, b)$ specifies the probability of finding a pixel of color $b$ at a distance $m$ from a pixel of color $a$ in the image. The color correlogram of an image is the probability of a joint occurrence of two pixels some distance apart, where one pixel belongs to a specific color and the other belongs to another. Each entry $(a, b)$ in the co-occurrence matrix expresses how many pixels whose color is $C_b$ can be found at a distance $d$ from $a$ pixel whose color is $C_a$. Each different value of $d$ leads to different co-occurrence matrix.

### 9.4.2 Shape-Based Image Retrieval

Shape is an important characteristic to identify objects. The term "shape" does not refer to the shape of an image, but to the shape of a particular region that is being sought out. Shapes will often be determined by first applying segmentation or edge detection to an image. Shape descriptors are classified into (i) boundary-based (or contour-based) and (ii) region-based methods.

This classification takes into account whether the shape features are extracted only from the contour or from the whole region. Subdivisions of structural (local) and global descriptors are based on whether the shape is represented as a whole or by segments/sections. Another possible classification categorizes shape description methods into spatial and transform domain techniques, depending on whether direct measurements of the shape are used or a transformation is applied.

#### 9.4.2.1 Moment Invariants

Each object is manifested by a 14-dimensional feature vector, along with two sets of normalized moment invariants: one from the object contour and another from its solid object silhouette. The Euclidean distance measures the similarity between different shapes represented by their moment invariants.

#### 9.4.2.2 Curvature Scale Space

The curvature scale space (CSS) descriptor represents a multiscale organization of the zero-crossing points of a planar curve for the MPEG-7 standard. A special matching algorithm is necessary to compare two CSS descriptors.

#### 9.4.2.3 Beam Angle Statistics

We can define a beam as the set of lines joining a contour pixel to the rest of the pixels along the contour. At each contour pixel, the angle between a pair of lines is calculated, and the shape descriptor is defined by using the third-order statistics of all the beam angles in a set of neighborhoods. The beam angle statistics (BAS) descriptor is based on the beams originating from a contour pixel. The similarity between two BAS moment functions is measured by an optimal correspondent subsequence algorithm.

#### 9.4.2.4 Tensor Scale Descriptor

In the tensor scale concept, a shape descriptor's tensor scale at any image point can be technically exhibited by the largest ellipse (2D) centered at that point, and is compared within a similar homogeneous region by using a correlation-based distance function. Tensor scale descriptor is acquired by extracting the tensor scale parameters of the original image and later creating the ellipse orientation histogram.

#### 9.4.2.5 Contour Salience

The contour salience (CS) descriptor comprises of the salience values of salient pixels and their location, and is used in a heuristic coordinated algorithm as a distance function. This method uses image foresting transform to compute the salience values of contour pixels for locating salient points along the contour to establish the relation between a contour pixel and its internal and external skeletons.

#### 9.4.2.6 Segment Salience

An improved version of the CS descriptor is segment salience, which incorporates two improved features, namely (i) the salience values of contour segments instead of salience values of isolated points, and (ii) matching algorithms that replace heuristic matching by an optimum approach. The salience values along with the contour are computed and the contour is divided into a predefined number of segments of equivalent size. The internal and external influence areas of each segment are computed by summing up the influential areas of their corresponding pixels.

#### *9.4.2.7 Distance Transforms*

Distance transform is another approach that includes shape information. The method takes a binary image of feature and nonfeature pixels and calculates the distance of every pixel with the closest feature. Although potentially expensive, this highly efficient algorithm has been developed, which requires two passes through the image to generate output.

### 9.4.3 Texture-Based Image Retrieval

This characteristic is represented by the basic image's existence, whose spatial distribution generates some visual patterns outlined in terms of directionality, granularity, and repetitiveness. Texture measures are used for visual patterns in images. Textures [15] are signified by texels that are formerly placed into a number of sets based on how many textures are detected in the image.

#### *9.4.3.1 Space-Based Approaches*

The space-based method identifies image regions that have a uniform texture. Small local regions or pixels are combined based on the similarity of texture. Regions with dissimilar textures are then considered as segmented regions. This technique has the benefit of having the regions' boundaries always closed. Consequently, the regions with different textures are always well separated. The co-occurrence matrix is a traditional approach for encoding texture information. It describes spatial relationships among gray levels in an image. A cell defined by the position $(i, j)$ in this matrix registers the probability that two pixels of gray levels $i$ and $j$ occur in two relative positions. Hence, a co-occurrence probability set is proposed to characterize textured regions.

#### *9.4.3.2 Frequency-Based Texture Descriptors*

This includes the Gabor wavelet coefficients. Apart from Gabor wavelet coefficients, previous works have shown that time frequency–based texture descriptors have also been a part of the CBIR feature extraction technique. In addition to Gabor features and gray-level co-occurrence matrix, texture feature coding method has been used in frequency-based texture descriptors. Often it has been seen that texture descriptors have been used in frequency domains in order to achieve the objective of the framework.

#### *9.4.3.3 Texture Signatures*

This descriptor aims to describe texture information in terms of coarseness, contrast, and directionality. With the intention of accomplishing this, it is required to have an a priori knowledge of the recognized classes.

### 9.4.4 Clustering-Based Image Retrieval

Data clustering is a technique for extracting hidden patterns from massive data sets, possibly with high dimensionality [8]. Clustering methods can be classified into supervised, which requires human interaction to generate splitting criteria, and unsupervised schemes.

A finite unlabeled data set is separated into a finite/discrete accurate characterization set of unobserved samples generated to reach clustering goals.

#### 9.4.4.1 Relevance Feedback

Due to lack of proper or appropriate keywords to describe an image, keyword-based image retrieval is hard to deal with. To overcome this problem, the "relevance feedback" technique is managed by utilizing user-level feedback, hence reducing possible errors and redundancy. A Bayesian classifier is used that deals with positive and negative feedback. Content-based clustering methods are static in nature, so they cannot adapt to frequent user changes.

#### 9.4.4.2 Log-Based Clustering

Clustering can be done on the basis of retrieval system logs, preserved by an information retrieval procedure where session keys are formed and accessed for retrieval. Each session cluster generates a log-based document according to which a log-based [9] vector is formed for each session vector. Replacement of a session cluster is conducted with this log-based vector. The unaccessed log documents create their own vector.

#### 9.4.4.3 Hierarchical Clustering

To organize data into a categorized structure based on the proximity matrix, hierarchical clustering algorithms are used. The results are depicted by a binary tree or dendrogram that indicates the patterns' nested grouping and similarity levels at which groupings change.

#### 9.4.4.4 Retrieval Dictionary-Based Clustering

A classification retrieval system is constructed by manipulating the distance between two learned patterns, classifying them into different clusters, followed by a retrieval stage as a drawback of distance calculation. To overcome this difficulty, a retrieval system is used containing a retrieval dictionary generation unit that classifies learned patterns into multiple clusters and forms a retrieval dictionary using the clusters.

#### 9.4.4.5 N-Cut Algorithm

The N-cut concept is to establish nodes into groups, where high similarity within the group and/or low similarity between the groups is achieved. This technique is implemented by trial and error, and has been shown to be relatively robust in image segmentation by repeated application for getting more than two clusters. In this technique, the subgraph with the maximum number of nodes is randomly selected each time and broken up to yield a tree till the N-cut value exceeds some threshold.

#### 9.4.4.6 K-Means Clustering

Depending on the final required number of clusters, this nonhierarchal technique originally takes the number of components of a population. On the ground of the final essential number, mutually farthest apart clusters are chosen and then each component [11] in the population is examined and assigned to one of the cluster's components at the smallest distance. Recalculation is done every time for the centroid position after adding a component to that cluster, and this persists till all the components are grouped into the final required number of clusters.

#### 9.4.4.7 Graph Theory-Based Clustering

The fundamental graph theory properties make it very handy to depict clustering difficulties. For a weighted graph node, related data points in the pattern space and edges reproduce the proximities between each data point pair to detect clusters of any size and shape without requiring the cluster's actual number.

#### 9.4.4.8 Divide and Conquer K-Means

For large datasets, the divide and conquer procedure first divides the whole dataset into a subset founded on some benchmarks and again these subsets are clustered with the help of the K-means clustering algorithm to accelerate the search and to reduce complexity.

## 9.5 Comparative Study

A comparative study of the methods discussed so far is presented in Table 9.1. Table 9.1 discusses the proposed methodology in the particular papers and their limitations.

## 9.6 Conclusion

CBIR is the most extensively used image-processing operation by researchers. The major application of CBIR is in image databases. Image retrieval from a large database is as important as other image-processing operations. Nowadays medical images have also started using this method in order to retrieve patient data robustly and quickly. Reducing complexity and obtaining images correctly is the foremost goal of CBIR framework. Feature extraction is as necessary as similarity measurement for CBIR. It has been discussed previously that the color feature is one of the most popular attributes for the feature extraction process in CBIR, although it has been proved that texture also plays a major role during feature extraction. High-level semantics and low-level semantics are both necessary for the CBIR system.

Overall, it can be said that each component of the framework plays a vital role in making the system more accurate, but to reduce complexity, certain attributes have to be chosen wisely according to the requirement. Optimization has been used previously to increase the accuracy of the framework by using local search and global search techniques. Typically, no extensive study has been conducted in the CBIR optimization domain, which keeps the research area open for future work. Only particle swarm optimization has been used to

**TABLE 9.1**

Related Works for Previous Methodologies and Limitations

| Sl. No. | Paper Title | Authors | Methodology | Limitation |
|---|---|---|---|---|
| 1 | Feature reweighting in content-based image retrieval | Das et al. (2006) | In this work different weight-based feature extraction techniques were discussed. | Feature re-weighting was discussed but how that method was related to low-level or high-level features wasn't discussed. |
| 2 | Content-based image retrieval system for multi-object images using combined features | Katare et al. (2007) | In this paper a gradient vector flow active contour-based shape segmentation technique was used on multiple objects. This method also used automated initialization for active contours. | Shape segmentation was used based on GVF, but the system's complexity wasn't discussed. |
| 3 | Feature selection for content-based image retrieval | Guldogan and Gabbouj (2008) | The proposed system's objectives were<br>(i) enhancing semantic image retrieval results<br>(ii) decreasing retrieval process complexity<br>(iii) improving the overall system usability | The system didn't discuss if similarity measurement was involved. |
| 4 | Content-based image retrieval methods | Vassilieva (2009) | A survey on the feature extraction and representation methods/metrics of the corresponding feature spaces was included. Color, shape, and texture features were considered. The authors provided detailed classification of the feature representations. | The paper didn't discuss the effectiveness of the methods included in the survey. |
| 5 | Content-based image retrieval using color and edge direction features | Zhang and Zou (2010) | For image retrieval, the color index codes along with the edge direction feature were used. The color feature's weight belonging to the same color feature's subblock was included. | Only used color as a feature. |
| 6 | Content-based image retrieval: some basics | Schaefer (2011) | The author depicted some of the main image features that can be used for CBIR. | Detailed analysis was not done. |
| 7 | Content-based image retrieval using various distance metrics | Patil and Talbar (2012) | For image retrieval, the authors compared six different distance metrics, namely the Canberra, Euclidean, Bray–Curtis, Manhattan, square chi-squared distances, and square chord to determine the best similarity measure. Furthermore, the pyramid structured wavelet decomposition and the energy levels were calculated. | Semantic image retrieval wasn't discussed. |

(Continued)

**TABLE 9.1** ***(Continued)***

Related Works for Previous Methodologies and Limitations

| Sl. No. | Paper Title | Authors | Methodology | Limitation |
|---|---|---|---|---|
| 8 | A novel approach for content-based image retrieval | Singh et al. (2012) | In a dynamic environment, the CBIR problem was addressed. The authors proposed a framework to select the most suitable features to analyze the newly received images to improve the retrieval efficiency and accuracy. The algorithm designed the feature vectors after segmentation that were employed in similarity comparison between the query image and database images. The proposed framework was trained for different images in the database. | The work didn't discuss how the efficiency was improved with the help of the newly proposed algorithm. |
| 9 | Using visual features to design a content-based image retrieval method optimized by particle swarm optimization algorithm | Chang et al. (2013) | A CBIR technique was proposed using three types of visual features and 12 distance measurements that were optimized by the PSO algorithm. | Other optimization algorithms and their results associated with CBIR weren't discussed. |
| 10 | Content-based image retrieval using color edge detection and discrete wavelet transform | Agarwal et al. (2014) | A novel algorithm for CBIR based on discrete wavelet transform (DWT) and color edge detection was proposed. The proposed method was different from the existing histogram-based approach. | Histogram analysis is not done. Grayscale components are not discussed. |
| 11 | Content-based image retrieval using the radon projections approach | Ghuge and Patil (2014) | A CBIR technique was implemented using conventional histogram and radon transform. This transform was based on image intensity projection along a radial line oriented at a specific angle. | Comparative study with other methods was not mentioned. |
| 12 | An efficient content-based image retrieval using EI classification and color features | Yasmin et al. (2014) | An effective technique for image search and retrieval was suggested in this study. Thus, images were decomposed into a minimum of 16 up to 24 squares of equal size, and later edge detection was applied to those decomposed parts. Pixel classification was performed on the basis of inner pixels and edge pixels. | Pixel classification region's justification wasn't provided. |

assess the framework's accuracy. Combining CBIR with other image-processing operations such as segmentation, clustering, image registration, and watermarking can also be an area to work on in the near future.

## References


1. A. Katare, S.K. Mitra, and A. Banerjee, Content based image retrieval system for multi object images using combined features, *International Conference on Computing: Theory and Applications (ICCTA '07)*, Kolkata, India, March 2007, pp. 595–599.
2. J. Zhang and W. Zou, Content-based image retrieval using color and edge direction features, *2010 Second International Conference on Advanced Computer Control (ICACC)*, Boston, MA, *Vol. 5*, March 2010, pp. 459–462.
3. S. Agarwal, A.K. Verma, and N. Dixit, Content based image retrieval using Color Edge detection and discrete wavelet transform, *2014 International Conference on Issues and Challenges in Intelligent Computing Techniques (ICICT)*, Ghaziabad, India, February 2014, pp. 368–372.
4. N.N. Ghuge and B.D. Patil, Content based image retrieval using Radon projections approach, in *ICT and Critical Infrastructure: Proceedings of the 48th Annual Convention of Computer Society of India—Vol. II*, Advances in Intelligent Systems and Computing, Springer, Vol. 249, 2014, pp. 145–153.
5. G. Schaefer, Content-based image retrieval: Some basics, in *Man-Machine Interactions 2*, Part 1, 2011, Springer, pp. 21–29.
6. E. Guldogan and M. Gabbouj, Feature selection for content-based image retrieval, *Signal, Image and Video Processing*, 2(3), September 2008, 241–250.
7. S. Patil and S. Talbar, Content based image retrieval using various distance metrics, in *Data Engineering and Management*, Lecture Notes in Computer Science, Vol. 6411, 2012, IGI Global, pp. 154–161.
8. G. Das, S. Ray, and C.L. Wilson, Feature re-weighting in content-based image retrieval, in *Image and Video Retrieval*, Lecture Notes in Computer Science, Vol. 4071, 2006, Springer, pp. 193–200.
9. H. Aboulmagd, N. El-Gayar, and H. Onsi, A new approach in content-based image retrieval using fuzzy, *Telecommunication Systems*, 40, February 2009, 55.
10. N.S. Vassilieva, Content-based image retrieval methods, *Programming and Computer Software*, 35(3), May 2009, 158–180.
11. M. Yasmin, M. Sharif, I. Irum, and S. Mohsin, An efficient content based image retrieval using EI classification and color features, *Journal of Applied Research and Technology (JART)*, 12(5), October 2014, 1–6.
12. B.-M. Chang, H.-H. Tsai, and W.-L. Chou, Using visual features to design a content-based image retrieval method optimized by particle swarm optimization algorithm, *Engineering Applications of Artificial Intelligence*, 26(10), November 2013, 2372–2382.
13. N. Singh, K. Singh, and A.K. Sinha, A novel approach for content based image retrieval, *Second International Conference on Computer, Communication, Control and Information Technology (C3IT-2012)*, February 25–26, 2012, Vol. 4, pp. 245–250.
14. A.K. Yadav, R. Roy, V. Yadav, and A.P. Kumar, Survey on content-based image retrieval and texture analysis with applications, *International Journal of Signal Processing, Image Processing and Pattern Recognition*, 7(6), 2014, 41–50.
15. R.D.S. Torres and A.X. Falcão, Content-based image retrieval: Theory and applications, *Revista de Informática Teórica e Aplicada*, 13, 2006, 161–185.
16. K. Vijay and Dr. R. Anitha, A content-based approach to image database retrieval, *Journal of Computer Applications*, 1(4), October–December 2008, 15–19.
17. A.N. Bhute and B.B. Meshram, Content based image indexing and retrieval, *International Journal of Graphics & Image Processing*, 3(4), November 2013, 235–247.

18. M. Jain and S.K. Singh, A survey on: Content based image retrieval systems using clustering techniques for large data sets, *International Journal of Managing Information Technology (IJMIT)*, 3(4), November 2011, 23–40.
19. S. Das, S. Garg, and G. Sahoo, Comparison of content based image retrieval systems using wavelet and curvelet transform, *The International Journal of Multimedia & Its Applications (IJMA)*, 4(4), August 2012, 137–155.
20. M. Subramanian and S. Sathappan, An efficient content based image retrieval using advanced filter approaches, *The International Arab Journal of Information Technology*, 12(3), May 2015, 229–237.
21. R.S. Choras, Content-based image retrieval—A survey, *Biometrics, Computer Security Systems and Artificial Intelligence Applications*, 2006, Vol. 3, No. 4, 31–45.
22. R. Datta, J. Li, and J.Z. Wang, Content-based image retrieval—Approaches and trends of the new age, *MIR'05*, Singapore, November 11–12, 2005, pp. 1–10.
23. N. Singhai and S.K. Shandilya, A survey on: Content based image retrieval systems, *International Journal of Computer Applications*, 4(2), July 2010, 22–27.
24. H. Müller, N. Michoux, D. Bandon, and A. Geissbuhler, A review of content-based image retrieval systems in medical applications—Clinical benefits and future directions, *International Journal of Medical Informatics*, 73, 2004, 1–23.

# 10

# *Knowledge Mining from Medical Images*

**Amira S. Ashour, Nilanjan Dey, and Suresh Chandra Satapathy**

**CONTENTS**



**ABSTRACT** Medical data mining is a significant process in knowledge discovery from medical databases. This chapter is essential for extracting convenient information from image data in medicine and the health sciences. Technology work as a cutting-edge basic in relevant areas is presented. This was done to fill the gap for evolving medical image databases instead of simply reviewing the present literature. This chapter initiates a discussion for the data mining and KDD context and their connection with other related domains. A detailed recent KDD real-world applications summary is offered. The data mining and KDD definitions are depicted, with common multistep KDD procedure defined. The chapter includes a variety of methodologies and related work in the medical domain applications for knowledge discovery. Furthermore, the chapter addresses numerous threads within their broad issues, including the KDD system requirements and data mining challenges. It was reported that several novel techniques are essential to develop the imaging databases for the next information infrastructure in biomedicine.

**KEY WORDS:** *data mining, medical image mining, medical image database, knowledge discovery.*

## 10.1 Introduction

Data mining combines database technology, statistics, data visualization, pattern recognition, machine learning, and expert systems. A database refers to a data collection that is organized for easily access its contents with management, and nay required update. It contains combinations of data files/records, where the database's manager can supply users with the required information, and has the capabilities to control, to access, to read, and to write specifying report for further analysis. Databases typically comprise query ability; thus, data mining methods are applied as an efficient technique for database queries.

Medical image databases serve the novel biomedical imaging technologies, as well as their clinical services applications, in research and education. Along the wide domains for applications, data are accumulated in a dramatic speed. Innovative computational theories, algorithms, tools, for extracting meaningful significant information (knowledge) from the digital data volumes are raised up. Such theories and algorithms become an emerging topic for knowledge discovery in databases (KDD) [1].

Generally, in the medical domain, data can be considered to be low-/high-level data. Data have many forms such as text, voice, image and/or multimedia. The low-level data are too massive for understanding and interpreting information for easy diagnoses, while the other data form is more compact as it includes medical reports, descriptive models of the process, and predictive models to estimate the future cases. At the abstract level, the KDD is oriented by developing several methods for constructing data sense. Thus, data mining techniques become the core process for pattern extraction and discovery from medical data [2].

In healthcare applications, physicians and specialists periodically evaluate the healthcare data analysis methods in a traditional way. The physicians then deliver reports to the sponsoring healthcare organization for future planning and decision making for healthcare management. This manual interpretation and analysis using the traditional method to transform the data into knowledge is inaccurate, highly subjective, expensive, and time consuming. However, with the rapid increase in the massive data amount and resources, manual data analysis becomes entirely unrealistic in all domains including healthcare [3–8]. Databases are increasing in size with the rise in number of objects/records and/or the rise in the number of attributes to an object. Thus, automated database information analysis becomes an urgent issue to provide more valuable and accurate knowledge discovery. In this regard the KDD is considered to be an endeavor to address the problem of handling the massive data amount in a digital form.

Navigation through information-rich databases becomes an innovative knowledge discovery challenge that requires intelligent agents. Health informatics is a quickly developing domain that is focused on relating information technology and computer science for health/medical data extraction and interpretation. It is the automation of health information in order to support the clinical care, training, administration of health services, and medical research to enhance health information processing by collecting, storing, effectively retrieving, and analyzing medical data for clinicians, administrators, and researchers [9–11]. Nevertheless, there is a deficiency in the efficient analysis methods for discovering the hidden knowledge from the gigantic healthcare databases.

Thus, data mining is known as a nontrivial extraction process of implied and impacted convenient information from the stored data in a database. It becomes the magic solution for KDD systems in various applications. The applied data mining methods to the medical data include unsupervised neural networks, support vector machine, apriori and FPGrowth, linear genetic programming, bayesian ying yang, association rule mining, naïve bayes and map (SOM), bayesian network, time series technique, clustering and classification, and decision tree algorithms [12–18].

The current chapter introduces the KDD process outlines and the main data mining approaches. This chapter signifies the common framework to offer a wide vision of the inclusive objectives and techniques used in the KDD. The organization of the remaining sections is as follows. Section 10.2 presents the concept of knowledge discovery in the databases. Section 10.3 reports the various medical image mining techniques in the KDD system. Searching in the medical image databases is a significant application for the KDD in the medical domain, introduced in Section 10.4. Various applications on knowledge discovery in medical databases are addressed in Section 10.5. The applications challenges and future aspects are discussed in Section 10.6. The chapter conclusion is summarized in Section 10.7.

## 10.2 Knowledge Discovery in Databases

Finding significant patterns in data can be called knowledge extraction, data mining, information harvesting/discovery, or data pattern processing. Data mining is concerned with data analysis for data in databases, and management information systems (MIS). Data mining applies precise algorithms to extract patterns from the data under concern. The supplementary stages of the KDD process include data preprocessing, data cleaning, data selection, prior knowledge fitting integration, and appropriate clarification of the mining outcomes. These KDD steps are indispensable to guarantee that convenient knowledge is imitative from the data. Blind use for data mining techniques can lead to the invalid and meaningless patterns discovery [19–22].

The KDD is the process of recognizing novel, potentially significant, valid, and ultimate patterns in data. Pattern extraction designates finding an appropriate (i) model for the data, (ii) structure from data, and (iii) providing high-level depiction of a dataset's contents. The KDD procedure consists of several steps including data preparation, knowledge evaluation, patterns searching, and refinement that are recurrent in several iterations [23]. Typically, pattern refers to knowledge in a popular view, while data mining in the KDD is the process of employing discovery algorithms and data analysis. Beneath suitable computational adeptness restrictions, a precise patterns enumeration over the data is produced through data mining. Applied computational limitations place severe constraints on the subspace discovered by the data mining procedure.

The progress in the KDD techniques is related to the improvement of databases, pattern recognition, machine learning, statistics, artificial intelligence, data visualization, knowledge acquisition, and computing approaches. The foremost objective in the large data sets context is the efficient extraction of high-level knowledge from low-level data. The data mining constituent of KDD relies on a variety of pattern recognition, machine learning, and statistics techniques to discover patterns from data through the data mining phase of the KDD procedure. The KDD procedure is considered to be a multidisciplinary process that incorporates methods beyond their scopes machine learning. It emphasizes the inclusive procedure of knowledge finding from data, the data storage and accessibility, handling huge datasets scaling, and interpreting/visualizing the results [24].

Artificial intelligence techniques related to KDD embrace machine discovery for machine discovery. Knowledge discovery from data is essentially an arithmetical challenge, where statistics affords a framework for measuring the vagueness during inferring of common patterns from a specific sample of an overall population [25,26]. Thus, data mining is an imperative process that based on the problem's statistical phases.

The databases can be considered the driving force for KDDs. For efficient data access, database techniques for ordering and grouping operations when accessing data, and adjusting queries are significant processes. Data warehousing is a popular development of assembling and data cleaning to be accessible for decision support and online analysis. It is included for data cleaning and data access. The data cleaning process is concerned with representing and handling missing data as well as handling errors and noise [27]. The data access is a well-defined technique for data retrieving and determining access routes to data that were complex to obtain as in the stored offline data.

Generally, the KDD process comprises the use of database through any compulsory selection, preprocessing, data transformations, and subsampling in the database. The data mining module of the KDD procedure is focused on the methods to extract patterns and to compute them from the data [28,29]. Figure 10.1 illustrates the inclusive KDD procedure for the interpretation and evaluation of the mined patterns to conclude the patterns for innovative knowledge. The KDD process contains significant iteration for its steps.

Figure 10.1 depicts that the KDD process [24,30], which is iterative, consists of several steps: (i) develop and introduce the data from the database along with the significant prior knowledge to recognize the aim of KDD procedure, (ii) select the dataset/data samples for producing a target dataset, (iii) data cleaning and preprocessing to remove noise and to handle missed information cases, (iv) transformation via data/variables reduction and projection to find the significant features representing the data based on the application

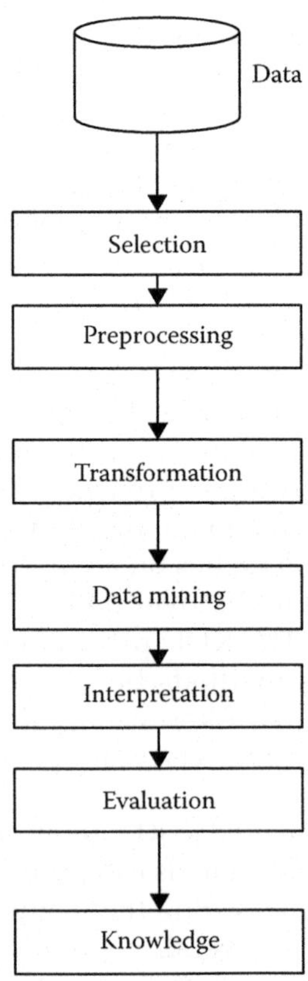


**FIGURE 10.1**
The KDD system steps.

under concern, thus invariant representation for the data is obtained, (v) matching the KDD procedure objectives to a specific data mining technique, which includes classification, summarization, clustering, and regression, (vi) use tentative analysis, models and hypothesis selection to determine the data patterns, (vii) interpreting the mined patterns by returning to the previous steps for further iteration and visualizing the extracted patterns/models/data, and (viii) evaluating the discovered knowledge by inspecting and determining the possible conflicts with formerly extracted knowledge.

## 10.3 Medical Image Mining Techniques in the KDD System

In medical data analysis and discovery, data mining techniques have sophisticated utility to handle such voluminous data. The data mining element of the KDD procedure includes repeated iterative application of a precise data mining technique. It is essential to use data mining procedures in order to assist decision support systems and predication in the healthcare domain. Medical data are stored in datasets for further analysis and transferring through numerous hospitals, clinics, and research centers. Nevertheless, most data mining applications in the clinical and decision support systems require centralized and homogeneous databases. Alternatively, data mining techniques can be realistic for scattered datasets to assist researchers in extracting perceptive patterns, designing predictive scoring systems, and finding the cause-and-effect relationships from the available data in the databases [31–35].

Knowledge discovery aims mainly for verification and discovery of the user's hypothesis and finding new patterns, respectively. In the discovery, the two main steps include the prediction and description of the patterns to a user. Data mining includes fitting models and patterns determination using the observed data. Most data mining techniques are based on pattern recognition, machine learning, and statistics including regression, clustering, and classification. Data mining procedures contain three main algorithmic mechanisms, namely, model representation, evaluation, and search.

There are numerous methods for data mining–based KDD method as follows:

- Regression [36] is a learning function for mapping data to a real-valued estimate variable, such as in estimating the probability of a patient's survival based on a set of diagnostic tests.
- Classification [37] is a learning function to classify data into some predefined classes such as the objects' automated identification in large image databases.
- Clustering [38–40] is a descriptive process to recognize a finite clusters (categories) set that is mutually exhaustive and exclusive and consists of a rich representation of the data.
- Summarization [41,42] is to discover an impact description for a data subset for automated report generation and interactive exploratory data analysis. Such methods include the functional relationships discovery between variables, summary rules derivation, and multivariate visualization methods.
- Dependency modeling [42,43] is concerned with finding a descriptive model for the significant dependencies between variables.
- Change and deviation detection [44,45] is used for realizing the most important data changes from formerly normative or measured values.

Recently, an extensive variety of data mining methods can be employed for medical image mining in the KDD systems, some of which are as follows.

- *Regression and classification methods* [46,47]: These methods are applied to predict the fitness of linear/nonlinear arrangements of basic functions to input variables arrangements. Such techniques comprise the adaptive spline approaches, feedforward neural networks, support vector machine, and projection pursuit regression.
- *Example-based methods* [48,49]: Case-based reasoning systems and nearest-neighbor classification and regression algorithms are examples of such techniques.
- *Decision trees and rules* [50–52]: These methods use the univariate splits to create easily the inferred model for the user to understand. Nevertheless, the limitation to a specific rule representation or tree can considerably control the functional form.
- *Probabilistic graphic dependency models* [53–55]: These methods use a graph structure to provide graphic models to specify the probabilistic dependencies. These models are applied with variables of discrete values within the framework of probabilistic expert systems.
- *Relational learning models* [56]: These methods employ the flexible pattern language of first-order logic. The main disadvantage of these approaches is the computational difficulties in terms of search.

## 10.4 Searching in Medical Image Databases

Image content is characterized by attributed interpersonal graphs that hold objects' features and relations between objects. In medical image databases, handling approximate search for data by using the image content is a critical process. Such techniques rely on the hypothesis that in a specified application domain, a fixed number of labeled (expected) objects (as lungs, heart, etc.) are common in all images. Furthermore, a variable number of the unlabeled objects (as tumor) also exist. The searching technique has the ability to answer queries. The stored images are recorded in a multidimensional space and are indexed using database approaches such as the R-trees.

Korn et al. [57] discussed the problem of finding similar tumor shapes in a medical database. The authors started from a natural similarity function, named the max morphological, to search for the nearest neighbors in huge assemblies of tumor-like shapes. The distance function was founded on signal processing approaches and mathematical morphology. The distance was invariant to translations and rotations. The feature index (F-index) approach was applied to handle the database for multimedia indexing. Petrakis and Faloutsos [58] proposed a technique that had several properties, namely, the database search is estimated, and where all images having a prespecified similarity degree is retrieved. In addition, no false dismissals exist.

Recently, the diagnostic procedure at hospitals is based mostly on comparing and reviewing images from various modalities and multiple time instances to screen the progression of diseases over a time period. Nevertheless, for vague cases, the physician and radiologist extremely rely on reference cases or second opinion. Even though there is a massive amount of attained medical images stored in the hospital systems that might be reclaimed for decision support, these data sets labor under weak search abilities. Consequently, Seifert et al. [59] proposed a search procedure to enable the doctor to achieve intellectual search on the medical image databases by linking appearance-based similarity search and ontology-based semantic search.

## 10.5 Applications on Knowledge Discovery in Medical Databases

The inclusive procedure of data mining and determining imperative knowledge from data is known as KDD. Data mining is a procedure of investigating huge data amounts in different perspectives to generate patterns that lead to intelligent systems. It plays an imperative role for knowledge discovery from historical data of various domains. In the healthcare area, data mining can be applied to mine the huge amount of medical data concerning diagnoses, diseases, patients, and medicine. This leads to improving the Quality of Service (QoS) through discovering hidden, potentially convenient features/patterns required for medical diagnosis [60].

Data mining has a significant role in medical applications such as medications, surgical procedures prediction, medical tests, and relationships between pathological data and clinical data discovery [31]. FPGrowth and Apriori are the most extensively used pattern mining algorithms [61] used in medical data mining. The authors in [15] applied unsupervised neural networks along with data visualization method for medical data mining. The authors depicted that the preprocessing step prior to medical data mining is a critical step. The authors in [62] established the essential use for data mining procedures to mine medical data content. They discussed the medical image mining challenges such as different medical coding systems used in hospital information systems, missing values in the databases, and data storage with respect to multivalued/temporal data.

In [63], symbolic rule extraction workbench for producing evolving rule sets was proposed. Extracting association rules from medical image data was proposed in [64]. The association rule mining realizes commonly stirring objects in the specified dataset. A Bayesian Ying Yang (BYY)–based classification technique that is a three-layered model was applied in [65]. A liver disease dataset was classified through automatic discovery of medical patterns. Islam et al. [16] employed the decision tree data mining procedure for lung cancer diagnosis image analysis using x-ray images. A classification system is proposed in Reference 17 to diagnose cardiovascular diseases.

Ghannad-Rezaie et al. [66] integrated a rule mining technique based on particle swarm optimization (PSO) technique to classify patients' dataset. The results revealed that the suggested technique accomplished surgery candidate selection procedure efficiently in epilepsy cases. Hassan and Verma [67] offered a hybrid methodology for medical data classification by combining K-means, self-organizing map (SOM), and Naïve Bayes with neural network classifier. All data were clustered in soft clusters by the neural network and the statistical clustering. Afterward, these clusters were fused using parallel and serial fusion in combination with the classifier. This method was tested and implemented on a benchmark medical database.

A classification model of diabetic database was proposed by Karegowda and Jayaram [68] using two schemes in cascading manner to develop the classification accuracy.

The suggested techniques were the genetic algorithm (GA) and the correlation-based feature selection (CFS). The correlation between the attributes decide the fitness of the individual that used in the mating. Thus, the GA attributes are reduced via the global search with fitness produced by the CFS. Afterward, a fine-tuned classification was completed using the neural networks classifier. The results indicated that the feature subset defined by the offered filter led to improved classification accuracy. Hogl et al. [69] presented a language named the knowledge discover question language for formulating questions to be used for knowledge discovery from medical data. The authors explored techniques for intelligent medical data mining that can be used for medical quality management.

## 10.6 Applications, Challenges, and Future Aspects

Researchers are interested in developing a variety of image analysis/processing techniques in the medical field for various applications [70–80] for efficient diagnoses. This leads to massive amount of data stored in datasets/databases that requires KDD systems for knowledge discovery. The applied criteria for KDD systems contain the absence of simpler alternate solutions, the probable impact of an application, and robust structural support for using technology. Privacy and legal issues are challenging tasks for applications handling personal data. Another technical criterion to be considered is the availability of adequate data, where more fields in the database require more compound patterns being required, leading to additional data requirements. Nevertheless, strong preceding knowledge can diminish the required number of cases. The relevance of attributes is another challenge that should be considered. It is imperative to have data attributes that are applicable to the discovery process.

Another challenge for the KDD process is the reduction of the noise levels, where high amounts of noise complicate the patterns identification process. Moreover, one of the immense significant concerns is prior knowledge about the application domain, the important relationships within the databases, knowing the user utility function, and the already known patterns. This prior knowledge assists the new knowledge discovery and patterns' identification.

In the field of data mining, feature subset selection is of great concern. The data's increased dimensionality makes classification training and testing very complex and time consuming. Thus, it is recommended to improve feature selection methods to support the KDD systems. Furthermore, in the data mining context, the development of artificial intelligence techniques as well as machine learning to support the KDD system is compulsory. The KDD places an extraordinary stress on finding comprehensible patterns for interpreting convenient knowledge. The KDD ought to emphasize robustness and scaling properties of modeling algorithms for noisy datasets.

In the future, it is recommended to develop techniques that concerning with large data volumes. Moreover, in the case of high dimensionality, because of the large number of fields (variables, attributes), the search space size increases in an explosive manner. This leads to spurious patterns extraction during the data mining process. Thus, dimensionality reduction methods and the prior use of knowledge to recognize unrelated variables are solutions for the preceding problem.

Solutions for the over-fitting are necessary, where during the search process for the paramount parameters for a specific model using an inadequate dataset, it may model any noise specific to the dataset along with the general patterns in the data. This leads to poor model performance in the test phase of the data. Probable solutions include sophisticated statistical strategies, cross-validation, and regularization. Nonstationary data can make formerly discovered patterns invalid. Such nonstationary cases may occur due to modifying, deleting, or augmenting the variables calculated in a specified application database can be with new calculations over time. Probable solutions can comprise incremental techniques for patterns updating and handling change by using it to prompt the search for patterns of the change only. Finally, machine learning and artificial intelligence techniques can possibly contribute considerably to several aspects of the KDD process.

## 10.7 Conclusions

Knowledge depiction contains new concepts for representing, accessing knowledge, and storing for signifying knowledge.

Data mining has a significant role in the KDD systems. It is a very imperative method to mine knowledge simply from data (text, image, video, multimedia, etc.). Data mining techniques can handle hidden knowledge extraction, data association, and supplementary patterns that are not clearly accumulated in the data. The most vital function of the mining is to produce all central patterns without preceding information of the patterns. Rule mining has been adjusting to massive medical image databases. Abundant researches have been carried out on this medical image mining to support the KDD systems in several applications such as the medical images knowledge discovery. This will lead to accurate diagnoses as more relevant information is extracted.

This chapter introduced a study on several image mining methods that were offered earlier in literature, where some definitions of basic concepts in the KDD system were introduced.

## References


1. Fayyad, U., G. Piatetsky-Shapiro, and P. Smyth. From data mining to knowledge discovery in databases. AI magazine. 1996 March 15;17(3):37.
2. Mortimore, W. C., D. A. Simon, and M. J. Gray. Computer based multimedia medical database management system and user interface. U.S. Patent 5,950,207, issued September 7, 1999.
3. Levine, A. B. Comparative medical-physical analysis. U.S. Patent 4,852,570, issued August 1, 1989.
4. Segal, E., M. Klein, and E. Kinchen. Method and system for managing patient medical records. U.S. Patent Application 09/776,673, filed February 6, 2001.
5. Shukla, D. P., S. B. Patel, and A. K. Sen. A literature review in health informatics using data mining techniques. *International Journal of Software and Hardware Research in Engineering* 2(2) (2014): 123–129.
6. Wennberg, D. Systems and methods for analysis of healthcare provider performance. U.S. Patent Application 11/542,574, filed October 3, 2006.

7. Doi, K. Current status and future potential of computer-aided diagnosis in medical imaging. *The British Journal of Radiology* (2014).
8. Abdelhak, M., S. Grostick, and M. A. Hanken. *Health Information: Management of a Strategic Resource*. Elsevier Health Sciences, 2014.
9. Coiera, E. *Guide to Health Informatics*. Boca Raton, FL: CRC Press, 2015.
10. Mantas, J., E. Ammenwerth, G. Demiris, A. Hasman, R. Haux, W. Hersh, E. Hovenga et al. Recommendations of the International Medical Informatics Association (IMIA) on education in biomedical and health informatics—First revision. *Acta Informatica Medica* 18(1) (2010): 4.
11. Rooksby, J., M. Rost, A. Morrison, and M. C. Chalmers. Personal tracking as lived informatics. In *Proceedings of the 32nd Annual ACM Conference on Human Factors in Computing Systems*, pp. 1163–1172. ACM, 2014.
12. Sunil, J. and R. C. Jain. A dynamic approach for frequent pattern mining using transposition of database. In *Communication Software and Networks, 2010. ICCSN'10. Second International Conference*, 2010 Feb 26, pp. 498–501. IEEE, 2010.
13. Nguyen, T.-T. An improved algorithm for frequent patterns mining problem. In *Computer Communication Control and Automation (3CA), 2010 International Symposium*, 2010 May 5 (Vol. 1, pp. 503–507). IEEE. pp. 503–507. IEEE, 2010.
14. Brameier, M. and W. Banzhaf. A comparison of linear genetic programming and neural networks in medical data mining. *IEEE Transactions on Evolutionary Computation* 5(1) (2001): 17–26.
15. Shalvi, D. and N. DeClaris. An unsupervised neural network approach to medical data mining techniques. In *The 1998 IEEE International Joint Conference on Neural Networks Proceedings. IEEE World Congress on Computational Intelligence*, Vol. 1, pp. 171–176. IEEE, 1998.
16. Islam, Md R., M. Chowdhury, and S. Khan. Medical image classification using an efficient data mining technique. In *Complex 2004: Proceedings of the Seventh Asia-Pacific Complex Systems Conference*, pp. 34–42. Central Queensland University, Rockhampton, Queensland, Australia, 2004.
17. Cheng, T.-H., C.-P. Wei, and V. S. Tseng. Feature selection for medical data mining: Comparisons of expert judgment and automatic approaches. In *19th IEEE Symposium on Computer-Based Medical Systems (CBMS'06)*, pp. 165–170. IEEE, 2006.
18. Tu, M. C., D. Shin, and D. Shin. A comparative study of medical data classification methods based on decision tree and bagging algorithms. In *Eighth IEEE International Conference on Dependable, Autonomic and Secure Computing (DASC'09)*, pp. 183–187. IEEE, 2009.
19. Piateski, G. and W. Frawley. *Knowledge Discovery in Databases*. MIT Press, 1991.
20. Fayyad, U., G. Piatetsky-Shapiro, and P. Smyth. From data mining to knowledge discovery in databases. *AI Magazine* 17(3) (1996): 37.
21. Frawley, W. J., G. Piatetsky-Shapiro, and C. J. Matheus. Knowledge discovery in databases: An overview. *AI Magazine* 13(3) (1992): 57.
22. Fayyad, U., G. Piatetsky-Shapiro, and P. Smyth. From data mining to knowledge discovery in databases. AI magazine. 1996 March 15;17(3):37.
23. Brachman, R. J. and T. Anand. The process of knowledge discovery in databases. In *Advances in Knowledge Discovery and Data Mining*, pp. 37–57. American Association for Artificial Intelligence, 1996.
24. Soibelman, L. and H. Kim. Data preparation process for construction knowledge generation through knowledge discovery in databases. *Journal of Computing in Civil Engineering* 16(1) (2002): 39–48.
25. Bankier, J. D., C. A. Beck, A. C. Brind, D. J. Brown, K. I. Brown, J. D. Burns, P. J. Docherty et al. Method and apparatus for knowledge discovery in databases. U.S. Patent 6,567,814, issued May 20, 2003.
26. Kanehisa, M., S. Goto, Y. Sato, M. Kawashima, M. Furumichi, and M. Tanabe. Data, information, knowledge and principle: Back to metabolism in KEGG. *Nucleic Acids Research* 42(D1) (2014): D199–D205.
27. Goebel, M. and L. Gruenwald. A survey of data mining and knowledge discovery software tools. *ACM SIGKDD Explorations Newsletter* 1(1) (1999): 20–33.
28. Fayyad, U., G. Piatetsky-Shapiro, and P. Smyth. The KDD process for extracting useful knowledge from volumes of data. *Communications of the ACM* 39(11) (1996): 27–34.

29. Casati, F., M.-C. Shan, and U. Dayal. Business processes based on a predictive model. U.S. Patent 7,565,304, issued July 21, 2009.
30. Rokach, L. and O. Maimon. *Data Mining with Decision Trees: Theory and Applications*. World Scientific, 2014.
31. Prather, J. C., D. F. Lobach, L. K. Goodwin, J. W. Hales, M. L. Hage, and W. Edward Hammond. Medical data mining: Knowledge discovery in a clinical data warehouse. In *Proceedings of the AMIA Annual Fall Symposium*, p. 101. American Medical Informatics Association, 1997.
32. Laurikkala, J., M. Juhola, E. Kentala, N. Lavrac, S. Miksch, and B. Kavsek. Informal identification of outliers in medical data. In *Fifth International Workshop on Intelligent Data Analysis in Medicine and Pharmacology*, pp. 20–24. 2000.
33. Loening, A. M. and S. S. Gambhir. AMIDE: A free software tool for multimodality medical image analysis. *Molecular Imaging* 2(3) (2003): 131–137.
34. Delen, D., G. Walker, and A. Kadam. Predicting breast cancer survivability: A comparison of three data mining methods. *Artificial Intelligence in Medicine* 34(2) (2005): 113–127.
35. Prokosch, H.-U. and T. Ganslandt. Perspectives for medical informatics. *Methods of Information in Medicine* 48(1) (2009): 38–44.
36. Rokach, L. and O. Maimon. *Data Mining with Decision Trees: Theory and Applications*. World Scientific, 2014.
37. Ahmed, A. B. E. D. and I. S. Elaraby. Data mining: A prediction for student's performance using classification method. *World Journal of Computer Application and Technology* 2(2) (2014): 43–47.
38. Yu, H., Z. Liu, and G. Wang. An automatic method to determine the number of clusters using decision-theoretic rough set. *International Journal of Approximate Reasoning* 55(1) (2014): 101–115.
39. Holzinger, A., M. Dehmer, and I. Jurisica. Knowledge discovery and interactive data mining in bioinformatics-state-of-the-art, future challenges and research directions. *BMC Bioinformatics* 15(6) (2014): 1.
40. Gupta, G. K. *Introduction to Data Mining with Case Studies*. PHI Learning Pvt. Ltd., 2014.
41. Tsai, C.-W., C.-F. Lai, M.-C. Chiang, and L. T. Yang. Data mining for internet of things: A survey. *IEEE Communications Surveys and Tutorials* 16(1) (2014): 77–97.
42. Zhang, L. and B. Liu. Aspect and entity extraction for opinion mining. In *Data Mining and Knowledge Discovery for Big Data*, pp. 1–40. Heidelberg, Germany: Springer, 2014.
43. Wu, X., X. Zhu, G.-Q. Wu, and W. Ding. Data mining with big data. *IEEE Transactions on Knowledge and Data Engineering* 26(1) (2014): 97–107.
44. Otten, S., M. Spruit, and R. Helms. Towards decision analytics in product portfolio management. *Decision Analytics* 2(1) (2015): 1.
45. Kasemsap, K. The role of data mining for business intelligence in knowledge management. *Integration of Data Mining in Business Intelligence Systems* (2015): 12–33.
46. Chaurasia, V. and S. Pal. Data mining techniques: To predict and resolve breast cancer survivability. *International Journal of Computer Science and Mobile Computing* 3(1) (2014): 10–22.
47. Cao, X., Y. Wei, F. Wen, and J. Sun. Face alignment by explicit shape regression. *International Journal of Computer Vision* 107(2) (2014): 177–190.
48. Shatkay, H. and R. Feldman. Mining the biomedical literature in the genomic era: An overview. *Journal of Computational Biology* 10(6) (2003): 821–855.
49. García, S., J. Luengo, and F. Herrera. *Data Preprocessing in Data Mining*. New York: Springer, 2015.
50. Agrawal, R., M. Mehta, and J. J. Rissanen. Data mining method and system for generating a decision tree classifier for data records based on a minimum description length (MDL) and presorting of records. U.S. Patent 5,787,274, issued July 28, 1998.
51. Yang, Q. and W. Xindong. 10 challenging problems in data mining research. *International Journal of Information Technology & Decision Making* 5(04) (2006): 597–604.
52. Peña-Ayala, A. Educational data mining: A survey and a data mining-based analysis of recent works. *Expert Systems with Applications* 41(4) (2014): 1432–1462.

53. Lafferty, J., A. McCallum, and F. Pereira. Conditional random fields: Probabilistic models for segmenting and labeling sequence data. In *Proceedings of the 18th International Conference on Machine Learning, ICML*, Vol. 1, pp. 282–289. 2001.
54. Dong, X., E. Gabrilovich, G. Heitz, W. Horn, N. Lao, K. Murphy, T. Strohmann, S. Sun, and W. Zhang. Knowledge vault: A web-scale approach to probabilistic knowledge fusion. In *Proceedings of the 20th ACM SIGKDD International Conference on Knowledge Discovery and Data Mining*, pp. 601–610. ACM, 2014.
55. Kimmig, A., L. Mihalkova, and L. Getoor. Lifted graphical models: A survey. *Machine Learning* 99(1) (2015): 1–45.
56. Tang, L. and H. Liu. Relational learning via latent social dimensions. In *Proceedings of the 15th ACM SIGKDD International Conference on Knowledge Discovery and Data Mining*, pp. 817–826. ACM, 2009.
57. Korn, F., N. Sidiropoulos, C. Faloutsos, E. Siegel, and Z. Protopapas. *Fast Nearest Neighbor Search in Medical Image Databases*. 1998.
58. Petrakis, E. G. M. and A. Faloutsos. Similarity searching in medical image databases. *IEEE Transactions on Knowledge and Data Engineering* 9(3) (1997): 435–447.
59. Seifert, S., M. Thoma, F. Stegmaier, M. Hammon, M. Kramer, M. Huber, H.-P. Kriegel, A. Cavallaro, and D. Comaniciu. Combined semantic and similarity search in medical image databases. In *SPIE Medical Imaging*, p. 796703. International Society for Optics and Photonics, 2011, pp. 796703–796703.
60. Ilayaraja, M. and T. Meyyappan. Mining medical data to identify frequent diseases using Apriori algorithm. In *2013 International Conference on Pattern Recognition, Informatics and Mobile Engineering (PRIME)*, pp. 194–199. IEEE, 2013.
61. Ma, H.-B., J. Zhang, Y.-J. Fan, and H. Yun-Fa. Mining frequent patterns based on IS+-tree. In *Proceedings of 2004 International Conference on Machine Learning and Cybernetics*, Vol. 2, pp. 1208–1213. IEEE, 2004.
62. Tsumoto, S. Problems with mining medical data. In *The 24th Annual International Computer Software and Applications Conference (COMPSAC 2000)*, pp. 467–468. IEEE, 2000.
63. Abidi, S. S. R. and K. M. Hoe. Symbolic exposition of medical data-sets: A data mining workbench to inductively derive data-defining symbolic rules. In *Proceedings of the 15th IEEE Symposium on Computer-Based Medical Systems (CBMS 2002)*, pp. 123–128. IEEE, 2002.
64. Olukunle, A. and S. Ehikioya. A fast algorithm for mining association rules in medical image data. In *Canadian Conference on Electrical and Computer Engineering (IEEE CCECE 2002)*, Vol. 2, pp. 1181–1187. IEEE, 2002.
65. Shim, J.-Y. and X. Lei. Medical data mining model for oriental medicine via BYY binary independent factor analysis. In *Proceedings of the 2003 International Symposium on Circuits and Systems (ISCAS'03)*, Vol. 5, p. V-717. IEEE, 2003.
66. Ghannad-Rezaie, M., H. Soltanain-Zadeh, M.-R. Siadat, and K. V. Elisevich. Medical data mining using particle swarm optimization for temporal lobe epilepsy. In *2006 IEEE International Conference on Evolutionary Computation*, pp. 761–768. IEEE, 2006.
67. ZahidHassan, S. and B. Verma. A hybrid data mining approach for knowledge extraction and classification in medical databases. In *Seventh International Conference on Intelligent Systems Design and Applications (ISDA 2007)*, pp. 503–510. IEEE, 2007.
68. Karegowda, A. G. and M. A. Jayaram. Cascading GA & CFS for feature subset selection in medical data mining. In *IEEE International Advance Computing Conference (IACC 2009)*, pp. 1428–1431. IEEE, 2009.
69. Hogl, O., M. Muller, H. Stoyan, and W. Stuhlinger. On supporting medical quality with intelligent data mining. In *Proceedings of the 34th Annual Hawaii International Conference on System Sciences*, 10pp. IEEE, 2001, pp. 1–10.
70. Roy, P., S. Goswami, S. Chakraborty, A. T. Azar, and N. Dey. Image segmentation using rough set theory: A review. *International Journal of Rough Sets and Data Analysis (IJRSDA)*, IGI Global. 1(2): 62–74.

71. Samanta, S., N. Dey, P. Das, S. Acharjee, and S. S. Chaudhuri. Multilevel threshold based gray scale image segmentation using cuckoo search. In *International Conference on Emerging Trends in Electrical, Communication and Information Technologies (ICECIT)*, December 12–23, 2012.
72. Pal, G., S. Acharjee, D. Rudrapaul, A. S. Ashour, and N. Dey. Video segmentation using minimum ratio similarity measurement. *International Journal of Image Mining (Inderscience)* 1(1): 87–110.
73. Bose, S., A. Mukherjee, S. C. Madhulika, S. Samanta, and N. Dey. Parallel image segmentation using multi-threading and K-means algorithm. In *2013 IEEE International Conference on Computational Intelligence and Computing Research (ICCIC)*, Madurai, India, December 26–28, 2013.
74. Dey, N. and A. Ashour eds. *Classification and Clustering in Biomedical Signal Processing*, Advances in Bioinformatics and Biomedical Engineering (ABBE) Book Series. IGI, 2016.
75. Karaa, W. B. A., A. S. Ashour, D. B. Sassi, P. Roy, N. Kausar, and N. Dey. MEDLINE text mining: An enhancement genetic algorithm based approach for document clustering. In *Applications of Intelligent Optimization in Biology and Medicine: Current Trends and Open Problems*, 2015.
76. Chakraborty, S., N. Dey, S. Samanta, A. S. Ashour, and V. E. Balas. Firefly algorithm for optimized non-rigid demons registration. In *Bio-Inspired Computation and Applications in Image Processing*, Yang, X. S. and J. P. Papa eds. 2016.
77. Fadlallah, S. A., A. S. Ashour, and N. Dey. Chapter 11: Advanced titanium surfaces and its alloys for orthopedic and dental applications based on digital SEM imaging analysis. In *Advanced Surface Engineering Materials*, Advanced Materials, Tiwari, A. ed. WILEY-Scrivener Publishing LLC, 2016.
78. Kotyk, T., N. Dey, A. S. Ashour, D. Balas-Timar, S. Chakraborty, A. S. Ashour, and J. M. R. S. Tavares. Measurement of the glomerulus diameter and Bowman's space width of renal albino rats. In *Computer Methods and Programs in Biomedicine*. Elsevier, 2016 Apr 30;126:143–153.
79. Saba, L., N. Dey, A. S. Ashour, S. Samanta, S. S. Nath, S. Chakraborty, J. Sanches, D. Kumar, R. T. Marinho, and J. S. Suri. Automated stratification of liver disease in ultrasound: An online accurate feature classification paradigm. In *Computer Methods and Programs in Biomedicine*. Elsevier, 2016 Jul 31;130:118–134.
80. Ahmed, S. S., N. Dey, A. S. Ashour, D. Sifaki-Pistolla, D. Bălas-Timar, and V. E. Balas. Effect of fuzzy partitioning in Crohn's disease classification: A neuro-fuzzy based approach. In *Medical & Biological Engineering & Computing*. Springer, 2016 Jan 1;55(1):101–115.

# 11

# *Segmentation for Medical Image Mining*

**Amira S. Ashour and Nilanjan Dey**

**CONTENTS**



**ABSTRACT** Image mining is a critical method for direct knowledge mining from image after image processing. It is an interdisciplinary domain that incorporates several methods such as computer vision, image processing, data mining, machine learning, artificial intelligence, and database. In addition, in image mining, segmentation is considered to be the main stage. Image mining is employed in hidden information extraction, image data association, and supplementary patterns that are not collected in the images. The most significant mining purpose is the generation of all relevant patterns without previous knowledge about those patterns. Mining is done along with the combined images groups and their associated data. The current work introduces the image mining concept in the medical domain. It presents a survey on several image segmentation methods that were suggested in earlier studies. The medical image mining for computer-aided diagnosis is discussed. Furthermore, the machine learning–based segmentation for medical image mining is depicted. Several related applications as well as the challenges and the future perspective are also illustrated.

**KEY WORDS:** *data mining, medical image mining, image segmentation, computer-aided diagnosis, machine learning.*

## 11.1 Introduction

In daily life, a massive amount of image data is produced in several domains, such as medical, sports, military, astronomy, and photographic images. Image mining is a procedure endeavor that attracts researchers and expertise for extracting the most significant image features in order to produce image patterns. It has numerous applications in various fields including image retrieval, computer vision, and pattern recognition [1]. Image mining approaches can be classified into one that (i) extracts images from image databases or any other source, or (ii) mines the images mixture pool and related alphanumeric data.

Progress in medical technology enlarged the information density for further imaging studies toward accurate diagnosis. Such studies are directed to achieve superior anatomical details using better spatial resolution facilities, evaluation of more subtle/microscopic structures by improved contrast resolution as well as enhanced temporal image acquisition rate [2]. Nevertheless, the potential improvement in the imaging-based diagnostic approaches may result in data overload during the medical information processing. This often manifests an increased computational time due to the usage of acquisition combinations, interpretation, and processing times; even this massive increase in data does not always provide improved diagnosis. This leads to the necessity for medical data mining procedures for extracting significant information from the massive amount of imaging data to guarantee an enhancement in the patient healthcare through accurate/early diagnosis with computational time reduction. This empowers physicians to spend less time interrelating with an image volume during clinical information extraction from these images.

Typically, image mining is not only for recovering relevant images but also for innovating patterns for the images that are remarkable. The foundation of an image mining system is considered to be a complicated process as it involves the integration of diverse techniques ranging from indexing schemes and image retrieval to data mining and pattern recognition to support the enriched diagnostic accuracy. Thus, image processing has an equally imperative role in image mining systems. Such image processing procedures include several phases, namely enhancement, features extraction/selection, segmentation, and classification [3–13]. Medical image retrieval, processing, matching, pattern recognition for extracting features like the color, shape, texture, and size from large image databases is a significant process. The required number of features for image representation can be massive [14]. Subsequently, using all accessible features for objects, recognition can lead to curse dimensionality. The image mining preprocessing phase includes feature extraction and selection.

Typical processing techniques apply segments to the image under concern, then characterize their features. Segmentation refers to the understanding as partitioning/decomposing the whole image into homogeneous parts/regions through boundary detection and texture analysis. Afterward, feature extraction is applied to find areas with definite properties, including edges, lines, corners as well as any set of measurements. These features are used for further pattern recognition and classification [15].

Computer-aided diagnosis (CAD) systems diagnose, for example, several abnormalities in tissues using the generated images from different medical imaging modalities [16]. Almost all the clinical imaging applications share a common approach for transforming raw imaging data to clinically relevant information using knowledge-based data mining algorithms. Recently, there has been a rapid progression in the medical image mining procedures for designing CAD systems to diagnose several diseases such as lung-, colon-, breast-, and brain-cancer. This attracts the researchers to offer the latest advancements in medical image mining for CAD system design.

The main objective of this chapter is to present an extensive overview of image segmentation–based medical image mining. The concepts for data mining, image mining, medical image segmentation, and machine learning–based segmentation for medical image mining are extensively discussed. The remaining sections are structured as follows. In Section 11.2, the medical image mining for computer-aided diagnosis is discussed.

Then, the segmentation techniques and generations are addressed in Section 11.3. The machine learning–based segmentation for medical image mining is introduced in Section 11.4. Section 11.5 presents several applications for the segmentation-based medical image mining. Section 11.6 reports the challenges and future perspectives, followed by conclusion in Section 11.7.

## 11.2 Medical Image Mining for Computer-Aided Diagnosis

In clinical practice, medical imaging enables noninvasive, rapid, and in vivo and in vitro visualization and human body quantitative assessment for prognosis and diagnosis. The CAD system is a computer program–based diagnosis for different abnormalities in different tissues using the captured images produced from various medical imaging modalities. Such modalities, including x-ray, computed tomography (CT), magnetic resonance imaging (MRI), ultrasound (US), single photon emission computed tomography (SPECT), and positron emission tomography (PET), produce medical images. Furthermore, the fusion of different image modalities that combines structural and functional properties in a single image is of great interest. This leads to an ever-increasing amount of medical images in hospitals, medical centers, and clinics that produce large image archives in the world [17].

These archives allow medical physicians, radiologists, other professionals to inspect the patients' images along with their attached diagnostic report and detailed description. It entails knowledge about the diseases' progression as well as therapy monitoring. The CAD is usually used as a second reader along with the radiologist to perform an interpretation of the image, followed by executing the CAD algorithm, and highlights the identified structures by the CAD algorithm for the radiologist to interpret the extracted features. Typically, the main objectives of the CAD include (i) automating the CAD processes to handle massive number of images, (ii) achieving accurate and fast results, and (iii) supporting faster communication for patient care using information technology.

Consequently, data mining methods are required with medical image analysis/processing as a foundation for intelligent CAD to realize the indicative and pathologic information embedded in the large-scale medical image archives. Definitely, data mining is the procedure of extracting convenient nontrivial information from massive data repositories. Such data can be in the form of text, image, video, or multimedia. Image mining can extract automatically and semantically significant patterns and hidden information from the images' data massive amount. There is an association between the image processing and data mining [18].

Image mining techniques have increased popularity in several application domains, such as medicine where a massive amount of data are accessible in the form of medical images. However, the image quality has a significant role when applying the image mining techniques to improve diagnosis accuracy, thereby increasing the survival rates. Information discovery from stored data in alphanumeric databases, including relational databases, is a focal point for data mining. Image data are considered to be nonstandard data that can also be available as an extensive image data collection. These image data can be mined in order to realize valuable and new information.

Image mining can be categorized into (i) domain-specific applications that include the most significant image features extraction into a suitable form for data mining such as

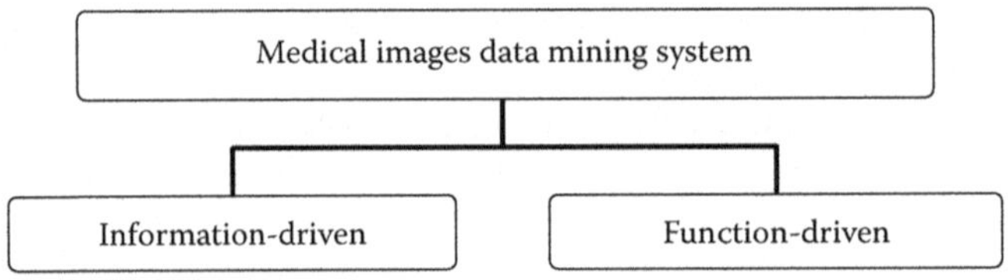


**FIGURE 11.1**
Medical images data mining system framework.

feature vectors [19–21] and (ii) general applications that generate image patterns for understanding the relations between the low-level image features and the image's high-level human perceptions [22–24]. Figure 11.1 illustrates the medical images data mining system frameworks, which can be either (i) information-driven, that is, designing a hierarchical construction with highlighting on the various hierarchy levels of the information [25] or (ii) function-driven, which serves to clarify and organize the various tasks to be executed in the image mining [26].

Information-driven framework has four information levels, namely pixel level, object level, pattern/knowledge level, and the semantic level. In medical image mining, preprocessing is considered the principal task to the image sets, which supports the data mining procedures. A feature vector of any image represents the relevant characteristics from the global set of features is based essentially on image segmentation. Various data mining algorithms are adapted for several tasks including association rule mining [27], image retrieval [28], segmentation and feature extraction [29], and classification [30] techniques. Advanced image mining techniques employ data mining approaches after preprocessing the image data to be proper for mining.

Generally, the image mining system includes content-based image retrieval approaches with image processing and databases. Consequently, the current chapter is interested with the segmentation for medical image data mining.

## 11.3 Segmentation Algorithms

Segmentation is defined as the separating/dividing procedure of an image into homogenous regions with similar characteristics including texture, color, gray level, contrast, and brightness [31]. The main roles of medical image segmentation are studying the anatomical structure, measuring the tissue volume that reflects the tumor growth, identifying the region of interest (ROI) to locate tumors and abnormalities, and assisting the treatment planning. The accessible techniques for medical images segmentation are explicit to the imaging modality, application, and specific body part to be investigated. Consequently, there is no general segmentation algorithm for all medical images, where each medical imaging system has its definite limitations.

Automatic medical images segmentation is a complicated process, where the medical images have a complex nature. Moreover, the segmentation algorithm output is affected by several factors, such as the intensity inhomogeneity, partial volume effect, gray level closeness of different soft tissues, and the artifacts produced by the used modalities. Typically, good segmentation is achieved when the pixels in the same region have similar

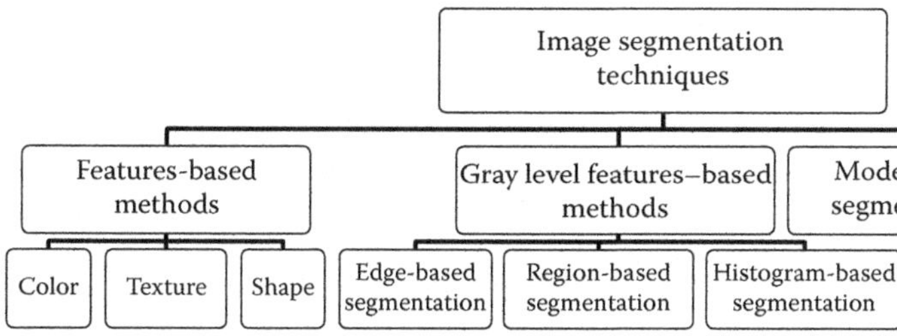


**FIGURE 11.2**
Image segmentation techniques.

gray scale of multivariate values and procedure a connected region, while the neighboring pixels from different regions have dissimilar values.

Medical image segmentation is a challenging process; thus, there are different categories of segmentation techniques [32–34]. Figure 11.2 demonstrates the wide-ranging categories for the segmentation techniques of an image, which are as follows: (i) approaches based on texture features [35], (ii) approaches based on gray level features [36], (iii) model-based segmentation [37], and (iv) atlas-based segmentation [38].

Figure 11.2 illustrated that the feature-based segmentation techniques segment the image based on texture, color, and/or shape features. These features are identified as low-level features. Texture feature in an image is a pattern of intensity variation formed by the different nature of the imaged object's surface, while the color feature is a point-like nature that makes them independent of the image size. Shape features of an image depend mainly on the shape of the objects/regions within an image.

In the model-based segmentation [37], the organ's structure has a repetitive geometry form that can be exhibited for deviation of geometry and shape. It involves appearance model and active shape, level-set-based models and deformable models. However, this segmentation technique suffers from the necessity of existing manual interaction to select the fitting parameters and to place an initial model. In addition, the original deformable models can provide deprived convergence to the concaved boundaries.

The most common powerful technique for medical image segmentation is the atlas-based segmentation approach which is considered to be one of the third-generation segmentation techniques [38]. In this technique, the size, shape, and different soft tissues/organs features are integrated in the form of a look-up table (LUT) or atlas. Nevertheless, it suffers from some limitations when segmenting complex structures with variable size, shape, and characteristics. Thus, professional knowledge is compulsory in constructing the database and for medical image mining.

Medical image segmentation can be categorized based on the generation (Figure 11.3) into:

1. *First generation* [39–41]: In this generation, segmentation is based on low-level techniques termed edge-based techniques, thresholding, and region-based techniques. Thresholding assigns pixels to groups (classes) based on the pixel's values range. Whereas, an edge filter is employed to the image in the edge-based segmentation, where pixels are categorized as nonedge or edge based on the filter output, and not disconnected pixels by an edge are considered to be from the same class. Moreover, region-based segmentation procedures work iteratively by combining together the neighbored pixels of similar values, while splitting pixels' groups having dissimilar value.

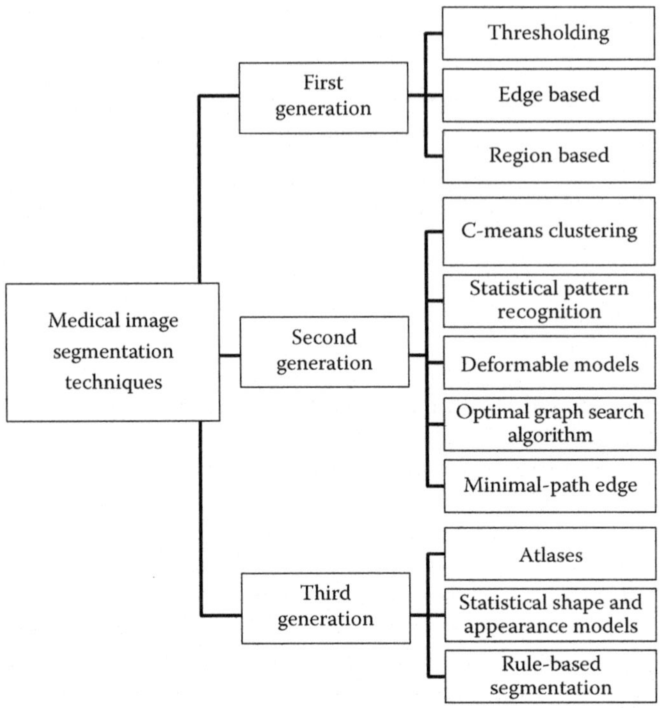


**FIGURE 11.3**
Medical images segmentation generations.

2. *Second generation* [42–45]: With the development of optimization techniques and uncertainty models, efforts are made to overcome the main first-generation segmentation problems in the second generation. This generation includes several techniques, namely, c-means clustering, statistical pattern recognition, deformable models, neural networks, optimal graph search algorithms, minimal-path edge following, multiscale analysis, and *statistically based* target tracking applied to the edge. However, these techniques remain data-dependent and cannot produce accurate and automatic segmentation in the general cases.
3. *Third generation* [46–47]: It incorporates higher-level information such as expert-defined rules, a priori information, and models of the desired object. Techniques used in this generation include atlases, tissue maps designed by manual segmentations and signified in a standard coordinate space to offer a priori information for statistical pattern recognition; *statistical shape and appearance models* impose limits on deformable models; and *rule-based segmentation*, where rules summarize the domain information, such as anatomical information of the position and shape of an image object.

Generally, the first generation employs the simplest image analysis techniques, the second generation is characterized by applying the optimization methods and the uncertainty models, and the third generation integrates information into the segmentation procedure toward fully automatic medical image segmentation. However, the first- and second-generation methods may require manual correction to be clinically accurate. The techniques in each generation can be related to the other techniques in the previous generations based on the boundary following, region identification, and pixel classification. Each technique has its advantages and disadvantages and based on the applied medical application, a specific segmentation method is employed. The segmentation process evaluation for in vivo images can be achieved by assessment with performed segmentations by experts, by the synthetic images analysis or images using datasets.

Furthermore, several parameters that are random in nature can be considered for the comparison of the different segmentation techniques. Such metrics used for the comparison between the different segmentation approaches include spatial information, region-continuity, computational complexity, speed, noise immunity, detection of multiple objects, peak signal to noise ratio, root mean square error, Mean Square Error (MSE), and the accuracy [48].

## 11.4 Machine Learning-Based Segmentation for Medical Image Mining

Automatically, medical image mining extracts semantically significant information from the medical image data. Medical image mining deals with hidden information extraction, image association, and data in the image databases. It attracts researchers and artificial intelligence expertise for content-based image retrieval, database, computer vision, data mining, digital image processing, and machine learning. Numerous machine learning procedures can be applied for accurate classification system toward patient diagnosis medical images with informative mathematical attributes. Such machine learning methods include the naive Bayesian classifiers, decision trees, K-nearest neighbors, Bayesian networks, and support vector machines [49].

Image mining framework with its tools for medical image analysis was presented by Perner [50]. In picture-archiving systems, the author provided a procedure for data mining. It determined the suitable knowledge for medical picture identification and examination from the database of images. Techniques were applied to obtain an attributes list for symbolic image descriptions. An expert labeled the images based on this list and stored descriptions in the database. The digital image processing was realized to achieve better imaging of precise image characteristics, or to attain expert-independent characteristic assessment. Decision-tree induction was applied to realize the expert knowledge in the database. The proposed data mining and image processing techniques were applied to Hep-2 cell-image segmentation, as well as in lung-nodule analysis for x-ray images, and MRI images for lymph-node analysis as well as for breast examination.

Massive information content is an important feature of any tissue microarray analysis (TMA) system. Since, tissue image mining is considered to be practical and fast, Gholap et al. [51] proposed a four-level system for content-based tissue image mining to exploit the pathologist's knowledge, pattern identification, and artificial intelligence. Information like color or disparity was exploited at the image processing and information levels, while, at the object level, the pathological objects including cell components were predicted. Afterward, at the semantic level, individual cells arrangement in a tissue image was inspected. At the highest (knowledge) level, the expert inspection was identified.

Image segmentation has a critical role in numerous medical imaging applications by automating the anatomical arrangements, explanation and supplementary ROI. In several medical images, automatic tumor recognition is significant for accurate handling for human life. Sheela and Shanthi [52] defined an image mining technique for segmentation and categorization of brain MRI data. The authors established a system for image mining methods to classify the images into normal or abnormal cases, and then divide the anomalous Brain MRI tissues to recognize the brain-related diseases.

Mueen et al. [53] suggested a new image classification approach based on multilevel image features and the support vector machine (SVM) learning procedure. For medical image classification, the proposed technique united the several local- or global- features that were used independently. Three features, namely, global levels, local, and pixel features, were attained and combined together in an entire feature vector. Thus, the principal component analysis (PCA) decreased the large dimensional feature vector. The experimental validation testing was performed to validate the proposed system's efficiency. The achieved recognition rate was 89%. Moreover, the proposed method was compared to the SVM classifier and the K-nearest neighbor (K-NN) for performance evaluation.

Machine learning techniques enable the interpretation of diagnostic images to increase the accuracy and reliability of the diagnostic process. Šajn and Kumar [54] introduced the long-term study results for using data mining methods and image processing in the medical imaging. Pre- and post-test possibilities, multiresolution feature extraction, texture depiction, feature construction, and data mining algorithms were carried out to improve the prognostic power of the clinical examination. This long-term study established three imperative milestones: (i) enhancement, (ii) more important enhancement using multiresolution image parametrization, and (iii) additional feature structure using the PCA, which provided higher accuracy level. Machine learning approaches in combination with the feature subset selection improved the diagnostic performance. Through this proposed approach the image features were determined and transformed the image from the matrix notation to a set of discrete (parameters)/numeric features that express valuable high-level (associated to the pixel intensities) information for discriminating between classes.

The structural description for diagnosis from the medical images has been applied as it has numerous good properties. Such properties include invariance to rotation and invariance to global brightness. They arrest statistical and structural information to identify the frequently occurring structures with the most discriminative characteristics.

Automatic illicit medicine pill retrieval and matching is a significant difficulty due to the increased circulation of the tablet-type illicit drugs. Thus, Lee et al. [55] proposed an automatic technique to match the pill images based on the appeared paths on the tablet to recognize the manufacturer and source of the illicit drugs. The extracted feature vector from tablet images is based on invariant moments and edge localization.

Enireddy and Reddi [56] proposed a system to retrieve diagnostic cases analogous to the query medical image for easy availability and content-based image retrieval (CBIR) of digital medical images that are stored in large databases. The Haar wavelet was applied for lossless image compression. Texture features and edges were extracted from the compressed medical images using Gabor transforms and Sobel edge detector; respectively. The support vector and naïve Bayes were applied for classification of retrieval. The medical images were stored in large databases for retrieving the diagnostic cases. The CBIR used algorithms for extracting appropriate features from the image to present a query image.

## 11.5 Segmentation-Based Medical Image Mining Applications

Generally, images signify thousands of words, while pixels do not have fitness. Thus, colored images with their brightness do not offer in-depth information. Some meaningful information can be processed with a single pixel; however, most of the image segmentation procedures have trouble while identifying the regions of interest. This problem occurs due to the variation in the image properties from one image to another. Thus, the segmentation process is the major process in complex medical images as it is based on the users' objective for the required information to be mined from an image. Since, recently, researchers, focus has changed from a binary labeling problem to a multilabel problem, segmentation-based medical image mining has become a significant research domain for accurate diagnosis and treatment.

Senthil et al. [57] proposed an image mining base level set segmentation for accurate brain tumor detection. An effective image segmentation method using Sushisen algorithms, image mining, and classification procedure were combined with fuzzy algorithm. Level set segmentation step and thresholding were applied to obtain accurate brain tumor discovery. The proposed technique can get benefits of the classification algorithms for image segmentation for minimal computation time. In addition, it can get advantages of the Fuzzy EM in the aspects of accuracy. The used image segmentation method's performance was evaluated through a comparison with some state-of-the-art segmentation algorithms in terms of processing time, accuracy, and performance. The accuracy was assessed by comparing the obtained results with the ground truth of each managed image.

Sickle cell disease is an inherited red blood cell disorder set that leads to abnormal hemoglobin on the patient's red blood cells. Normal hemoglobin has a disc shape, while the abnormal hemoglobin has crescent-/sickle-shape. Consequently, Revathi and Jeevitha [58] applied the watershed transformation based on region processing segmentation to identify the blood samples' boundary through finding the regional minima on the real images. The authors used an elliptical matching approach to identify the elliptical and circular objects in the cellular cluster having normal and elongated erythrocytes in the sickle cell disease image samples. In the blood smear samples, the segmentation was employed to identify the cells/cluster of the overlapping cells. After segmentation process, the borders of the segmented objects were obtained by using an automatic technique based on the gray scale intensities. Afterward, the circumference adjustment algorithm was applied to analyze the circular objects to detect the best fit to the arch. The watershed transformation procedure provided a global segmentation and border closure with high accuracy.

Androutsos et al. [59] implemented a recursive *HSV*-space segmentation system for identifying perceptually prominent color regions within an image. Dubey [60] designated an image mining procedure based on the color histogram, and texture of the image under concern. The query image was reserved, and then the color histogram and texture was formed to produce the resultant image. The authors investigated the histogram-based search methods and the color texture methods in two diverse color spaces, namely, the RGB (red-green-blue) and HSV (hue, saturation, value). The histogram search was able to discriminate an image using its color distribution. It was established that color distribution has a significant role in image retrieval.

## 11.6 Challenges and Future Perspectives

Image segmentation denotes the image partitioning into mutually exclusive, non-overlapped, and homogenous regions. In any medical image, segmentation is considered to be the most vital and crucial process for enabling the characterization, delineation, and visualization of regions of interest. Despite the continued exhaustive research, segmentation is still a challenge due to the varied image content, occlusion, nonuniform object texture, cluttered objects, image noise, and other factors, especially in the medical domain applications due to the different modalities.

Essentially, image segmentation can be semiinteractive or fully automatic. Thus, it is required to develop segmentation algorithms for both categories. In some scenarios, manual interaction–based segmentation may be error-prone (e.g., in seed selection case), while the fully automated method can provide error output (e.g., in watershed segmentation case), and in some cases interactive techniques are time consuming and laborious. Hence, it is practically unachievable to find a single segmentation technique to segment all variety of images. Prior knowledge on the image can provide superior results, which direct the focus toward segmentation-based image mining techniques to extract more valuable significant information from a massive amount of images or image databases.

There are numerous techniques and algorithms available for image segmentation; however, there is still the urgent need to develop a fast and efficient technique for medical image segmentation. In addition, to date there is no universally accepted procedure for image segmentation that can be applied for almost all applications. This is owing to several factors, including the spatial characteristics of the image continuity, the images, homogeneity, texture, partial volume effect, and the image content. Consequently, there is no unique technique that can be considered virtuous for neither all images type nor all techniques equally good for a particular type of image. Due to the entire overhead factors, medical image segmentation has a promising future as the universal segmentation algorithm is still remains as a challenging problem in the medical image processing world, especially where several mining, classification, and image retrieval techniques are based mainly on the image segmentation.

Another future scope can be directed to use different optimization-based meta-heuristic algorithms to optimize the parameters used in the different segmentation algorithms in order to improve their accuracies.

Several segmentation techniques–based medical image mining methods proposed definitely save much time and offer convincing results with discussion and interpretation by the physician. Such methods have imperative potential for biomedical imaging analysis as well as medical images data mining which can be preserved for noises added for future analysis and in future. In addition, improving the computational speed of segmentation methods as well as improving the accuracy of cell segmentation is a critical challenge for researchers.

Other associated research subject, also essential to determine meaningful discovered image patterns, can be addressed as follows: (i) find a scheme to signify the image pattern, so the spatial information, contextual information, and imperative image characteristics are reserved in the representation scheme, (ii) the way to represent an image pattern is a critical issue, (iii) the features selection techniques can be improved to select the features to be included further in the mining process, and (iv) image pattern visualization requires researches to recognize the mined patterns to the user in a visually-rich form.

## 11.7 Conclusion

Image mining refers to mining knowledge definitively from an image. Data mining that is employed in the image processing domain is known as image mining. Image segmentation is the foremost process for image mining. It order to extract, control, and find the hidden knowledge within an image, image information association with extra patterns that are not perceptibly gathered in images become interesting. Image mining incorporates some procedures such as data mining, image processing, computer vision, machine learning, and artificial intelligence. One of the most vital processes of mining is to create all main patterns without the necessity of prior knowledge of the patterns. Mining is treated according to the collection of images and their related data.

Since segmentation has a significant role in image mining, this chapter has reported the three generations that can be used in the segmentation process. To measure the segmentation quality, the similarity between the elements of the same region is measured as it should be similar with clear difference between elements of the other regions.

The segmentation process can be categorized based on the selected parameter for segmentation, such as the homogeneity, pixel intensity, discontinuity, topology, and the cluster data. Each approach has its own advantages and disadvantages. The result achieved by one method may not be the same as compared with an other method. The segmentation methods are particular to specific applications that can often achieve better performance. Selecting a suitable approach for the segmentation problem can be a complex dilemma, and, thus, the chapter has reported several challenges and the future scope.

## References


1. T.Y. Gajjar and N.C. Chauhan, A review on image mining frameworks and techniques, *International Journal of Computer Science and Information Technologies*, 3(3), 4064–4066, 2012.
2. J. Marotti, S. Heger, J. Tinschert, P. Tortamano, F. Chuembou, K. Radermacher, and S. Wolfart, Recent advances of ultrasound imaging in dentistry—A review of the literature, *Oral Surgery, Oral Medicine, Oral Pathology and Oral Radiology*, 115(6), 819–832, 2013.
3. P. Roy, S. Goswami, S. Chakraborty, A.T. Azar, and N. Dey, Image segmentation using rough set theory: A review, In: *Medical Imaging: Concepts, Methodologies, Tools, and Applications*, IGI Global, pp. 1414–1426, 2017.
4. G. Pal, S. Acharjee, D. Rudrapaul, A.S. Ashour, and N. Dey, Video segmentation using minimum ratio similarity measurement, *International Journal of Image Mining (Inderscience)*, 1(1), 87–110, 2015.
5. S. Samanta, N. Dey, P. Das, S. Acharjee, and S.S. Chaudhuri, Multilevel threshold based gray scale image segmentation using cuckoo search, in *International Conference on Emerging Trends in Electrical, Communication and Information Technologies (ICECIT)*, Anantapur, Andhra Pradesh, India, December 12–23, 2012.
6. S. Bose, A. Mukherjee, S. Madhulika Chakraborty, S. Samanta, and N. Dey, Parallel image segmentation using multi-threading and K-means algorithm, in *2013 IEEE International Conference on Computational Intelligence and Computing Research (ICCIC)*, Madurai, India, December 26–28, 2013.
7. N. Dey and A. Ashour, eds. *Classification and Clustering in Biomedical Signal Processing*, Advances in Bioinformatics and Biomedical Engineering (ABBE) Book Series, IGI, 2016.

8. W.B.A. Karaa, A.S. Ashour, D.B. Sassi, P. Roy, N. Kausar, and N. Dey, MEDLINE text mining: An enhancement genetic algorithm based approach for document clustering, *Applications of Intelligent Optimization in Biology and Medicine*, pp. 267–287, Springer International Publishing, 2016.
9. S. Chakraborty, N. Dey, S. Samanta, A.S. Ashour, and V.E. Balas, Firefly algorithm for optimized non-rigid demons registration, In: *Bio-Inspired Computation and Applications in Image Processing*, X.S. Yang and J.P. Papa, eds., 2016, Springer.
10. S.A. Fadlallah, A.S. Ashour, and N. Dey, Chapter 11: Advanced titanium surfaces and its alloys for orthopedic and dental applications based on digital SEM imaging analysis, *Advanced Surface Engineering Materials*, A. Tiwari ed., Advanced Materials, WILEY-Scrivener Publishing LLC.
11. T. Kotyk, N. Dey, A.S. Ashour, D. Balas-Timar, S. Chakraborty, A.S. Ashour, and J.M.R.S. Tavares, Measurement of the glomerulus diameter and Bowman's space width of renal albino rats, *Computer Methods and Programs in Biomedicine*, 126, 143–153, 2016.
12. L. Saba, N. Dey, A.S. Ashour, S. Samanta, S.S. Nath, S. Chakraborty, J. Sanches, D. Kumar, R.T. Marinho, and J.S. Suri, Automated stratification of liver disease in ultrasound: An online accurate feature classification paradigm, *Computer Methods and Programs in Biomedicine*, Elsevier, New York, 2016.
13. S.S. Ahmed, N. Dey, A.S. Ashour, D. Sifaki-Pistolla, D. Bălas-Timar, and V.E. Balas, Effect of fuzzy partitioning in Crohn's disease classification: A neuro-fuzzy based approach, *Medical & Biological Engineering & Computing*, 55(1), 101–115, 2017.
14. J. Sklansky, Image segmentation and feature extraction, *IEEE Transactions on Systems, Man, and Cybernetics*, 8(4), 237–247, 1978.
15. A. Wroblewska, P. Boninski, A. Przelaskowski, and M. Kazubek, Segmentation and feature extraction for reliable classification of microcalcifications in digital mammograms, *Optoelectronics Review*, 3, 227–236, 2003.
16. K. Doi, Computer-aided diagnosis in medical imaging: Historical review, current status and future potential, *Computerized Medical Imaging and Graphics*, 31(4), 198–211, 2007.
17. M.-L. Antonie, O.R. Zaiane, and A. Coman, Application of data mining techniques for medical image classification, in *MDMKDD'01 Proceedings of the Second International Conference on Multimedia Data Mining*, pp. 94–101, 2001.
18. C. Ordonez and E. Omiecinski, Discovering association rules based on image content, in *Proceedings of the IEEE Advances in Digital Libraries Conference (ADL'99)*, pp. 38–49, 1999.
19. U.M. Fayyad, S.G. Djorgovski, and N. Weir, Automating the analysis and cataloging of sky surveys, *Advances in Knowledge Discovery and Data Mining*, 471–493, 1996.
20. W. Hsu, M.L. Lee, and K.G. Goh, Image mining in IRIS: Integrated retinal information system, in *ACM SIGMOD*, 2000.
21. A. Kitamoto, Data mining for typhoon image collection, in *Second International Workshop on Multimedia Data Mining (MDM/KDD'2001)*, 2001.
22. W. Hsu, M.L. Lee, and J. Zhang, Mining: Trends and developments, *Journal of Intelligent Information Systems*, 19(1), 7–23, 2002.
23. O.R. Zaiane, J. Han, Z.N. Li, J.Y. Chiang, and S. Chee, MultiMediaMiner: A system prototype for multimedia data mining, in *Proceedings of ACM-SIGMOD*, Seattle, WA, 1998.
24. C. Ordonez and E. Omiecinski, Discovering association rules based on image content, in *IEEE Advances in Digital Libraries Conference*, 1999.
25. J. Zhang, W. Hsu, and M.L. Lee, An information-driven framework for image mining, in *12th International Conference on Database and Expert Systems Applications*, 2001.
26. J. Li and R.M. Narayanan, Integrated information mining and image retrieval in remote sensing, Chapter 16. In: C.I. Chang (ed.), *Recent Advances in Hyperspectral Signal and Image Processing*, 1st edn., Transworld Research Network, pp. 449–478, 2006.
27. A.J.T. Lee, R.-W. Hong, W.-M. Ko, W.-K. Tsao, and H.-H. Lin, Mining spatial association rules in image databases, *Information Sciences*, 177(7), 1593–1608, 2007.
28. K.L. Tan, B.C. Ooi, and C.Y. Yee, An evaluation of color-spatial retrieval techniques for large image databases, *Multimedia Tools and Applications*, 14(1), 55–78, Kluwer Academic Publishers, Dordrecht, the Netherlands, 2001.

29. K. Fukuda and P.A. Pearson, Data mining and image segmentation approaches for classifying defoliation in aerial forest imagery, PhD disseration, International Environmental Modelling and Software Society, 2006.
30. A. Vailaya, A.T. Figueiredo, A.K. Jain, and H.J. Zhang, Image classification for content-based indexing, *IEEE Transactions on Image Processing*, 10(1), 117–130, January 2001.
31. R.C. Gonzalez and R.E. Woods, *Digital Image Processing*, 2nd edn., Pearson Education, 2004.
32. R. Popilock, K. Sandrasagaren, L. Harris, and K.A. Kaser, CT artifact recognition for the nuclear technologist, *Journal of Nuclear Medicine Technology*, 36, 79–81, 2008.
33. D.L. Pham, C. Xu, and J.L. Prince, Current methods in medical image segmentation, *Annual Review of Biomedical Engineering*, 2, 315–337, 2000.
34. J.L. Prince and J.M. Links, *Medical Imaging Signals and System*, Pearson Education, 2006.
35. N. Sharma, A.K. Ray, S. Sharma, K.K. Shukla, S. Pradhan, and L.M. Aggarwal, Segmentation and classification of medical images using texture-primitive features: Application of BAM-type artificial neural network, *Journal of Medical Physics*, 33, 119–126, 2008.
36. N. Sharma and A.K. Ray, Computer aided segmentation of medical images based on hybridized approach of edge and region based techniques, in *Proceedings of International Conference on Mathematical Biology*. Mathematical Biology Recent Trends, Anamaya Publishers, pp. 150–155, 2006.
37. O. Ecabert, J. Peters, H. Schramm, C. Lorenz, J. von Berg, M.J. Walker, M. Vembar et al., Automatic model-based segmentation of the heart in CT images, *IEEE Transactions on Medical Imaging*, 27(9), 1189–1201, 2008.
38. P. Aljabar, R.A. Heckemann, A. Hammers, J.V. Hajnal, and D. Rueckert, Multi-atlas based segmentation of brain images: Atlas selection and its effect on accuracy, *Neuroimage*, 46(3), 726–738, 2009.
39. D.L. Pham, C. Xu, and J.L. Prince, A survey of current methods in medical image segmentation, Technical report, The Johns Hopkins University, Department of Electrical and Computer Engineering, Baltimore, MD, 1998.
40. D.D. Patil and S.G. Deore, Medical image segmentation: A review, *International Journal of Computer Science and Mobile Computing*, 2(1), 22–27, 2013.
41. M. Singh and A. Misal, A survey paper on various visual image segmentation techniques, *International Journal of Computer Science and Management Research*, 2(1), 1282–1288, 2013.
42. A. Funmilola, Fuzzy k-c-means clustering algorithm for medical image segmentation, *Journal of Information Engineering and Applications*, 2(6), 21–32, 2012.
43. S. Murugavalli and V. Rajamani, An improved implementation of brain tumor detection using segmentation based on neuro fuzzy technique, *Journal of Computer Science*, 3(11), 841–846, 2007.
44. H. Costin, A fuzzy rules-based segmentation method for medical images analysis, *International Journal of Computer Communication & Control*, 8(2), 196–205, 2013.
45. D. Jayadevappa, S.S. Kumar, and D.S. Murty, Medical image segmentation algorithms using deformable models: A review, *Institution of Electronics and Telecommunication Engineers (IETE)*, 28(3), 248–255, 2011.
46. N. Sharma and I.M. Aggarwal, Automated medical image segmentation technique, *Journal of Medical Physics*, 35(1), 3–14, 2010.
47. D. García-Lorenzo, S. Francis, S. Narayanan, D.L. Arnold, and D.L. Collins, Review of automatic segmentation methods of multiple sclerosis white matter lesions on conventional magnetic resonance imaging, *Medical Image Analysis*, 17(1), 1–18, 2013.
48. A.M. Khan and S. Ravi, Image segmentation methods: A comparative study, *International Journal of Soft Computing and Engineering (IJSCE)*, 3(4), 2231–2307, 2013.
49. Y. Peng, B. Yao, and J. Jiang, Knowledge-discovery incorporated evolutionary search for microcalcification detection in breast cancer diagnosis, *Artificial Intelligence in Medicine*, 37(1), 43–53, 2006.
50. P. Perner, Image mining: Issues, framework, a generic tool and its application to medicalimage diagnosis, *Engineering Applications of Artificial Intelligence*, 15(2), 205–216, 2002.
51. A. Gholap, G. Naik, A. Joshi and C.V.K. Rao, Content-based tissue image mining, in *IEEE Computational Systems Bioinformatics Conference (CSBW'05)*, pp. 359–363, 2005.

52. L. Jaba Sheela and V. Shanthi, Image mining techniques for classification and segmentation of brain MRI data, *Journal of Theoretical and Applied Information Technology*, 3(4), 115–121, 2007.
53. A. Mueen, M. Sapian Baba, and R. Zainuddin, Multilevel feature extraction and x-ray image classification, *Journal of Applied Sciences*, 7(8), 1224–1229, 2007.
54. L. Šajn and M. Kukar, Image processing and machine learning for fully automated probabilistic evaluation of medical images, *Computer Methods and Programs in Biomedicine*, 104(3), e75–e86, 2011.
55. Y.-B., Lee, U. Park, A.K. Jain, and S.-W. Lee, Pill-ID: Matching and retrieval of drug pill images, *Pattern Recognition Letters*, 33(7), 904–910, 2012.
56. V. Enireddy and K.K. Reddi, A data mining approach for compressed medical image retrieval, *International Journal of Computer Applications (0975–887)*, 52(5), August 2012.
57. P. Senthil, Image mining base level set segmentation stages to provide an accurate brain tumor detection, *International Journal of Engineering Science and Computing*, 6(7), 2016.
58. T. Revathi and S. Jeevitha, Efficient watershed based red blood cell segmentation from digital images in sickle cell disease, *International Journal of Scientific Engineering and Applied Science (IJSEAS)*, 2(4), April 2016.
59. D. Androutsos, K.N. Plataniotis, and A.N. Venetsanopoulos, A novel vector-based approach to color image retrieval using a vector angular-based distance measure, *Computer Vision and Image Understanding*, 75(1/2), 46–58, July/August 1999.
60. R.S. Dubey, Image mining using content based image retrieval system, *International Journal on Computer Science and Engineering (IJCSE)*, 02(07), 2353–2356, 2010.

# 12

# *Biological Data Mining: Techniques and Applications*

**Amira S. Ashour, Nilanjan Dey, and Dac-Nhuong Le**

CONTENTS



**ABSTRACT** The recent years have seen an exponential growth in the amount of biological information, including that on DNA (deoxyribonucleic acid) and protein sequences, which is accessible in open databases. This was supported by more attention to improve computational procedures to automatically classify large capacities of massive sequence data into several groups analogous to their structure, their role in the chromosomes, and/or their function. Broadly used sequence classification procedures were developed for modeling sequences in a way that traditional machine learning procedures, including neural network and support vector machines, can be employed easily. Furthermore, conventional data analysis methods often fail to handle huge data amounts professionally. In this context, data mining tools can be applied for knowledge extraction from large data amounts. Lately, the biological data collection such as DNA-/protein-sequences is increasing rapidly due to the development of current technologies and the exploration of new methods such as the microarrays. Consequently, data mining method is applied to extract significant information from the massive biological data sequences amount. One significant research area is the protein sequences classification into several classes/subclasses, or families. The current chapter provides a comprehensive coverage of data mining for biological sequences concept and applications. It includes related work of data mining biological applications with both fundamental concepts and innovative methods. Significant insight and suggested future research areas for biological data mining are introduced.

**KEY WORDS:** *data mining, bioinformatics tools, protein sequence analysis, biological data mining.*

## 12.1 Introduction

In the clinical context, biologists are speeding up their efforts to determine the underlying disease pathways through the understanding of biological processes. This leads to a clinical and biological data flood from DNA (deoxyribonucleic acid) microarrays, genomic/protein sequences, biomedical images, protein interactions, to electronic health records, and disease pathways. Ultimate data analysis is applied to exploit these data in order to discover new knowledge that can be interpreted into clinical applications. Challenges facing biologists in the post-genome epoch include management of incomplete/noisy data, integrating several data sources, and processing compute-intensive tasks. Data mining techniques are designed to manage such data analysis difficulties and to enable biologists and scientists employing data mining obtain meaningful discoveries/observations from the massive biological data in real-world applications [1].

Data mining is effectively included in various sectors, such as medical, finance, marketing, retail, and business. It is the procedure for searching correlations, patterns, and trends in large data volumes. Furthermore, it lies at the interface of database technology, statistics, data visualization, pattern recognition, expert systems, and machine learning. Databases are an assembly of organized data where its contents can definitely be retrieved, updated, and managed. In biological science, huge information amount is presented within biological data comprising rich information but having poor knowledge. Biological data comprises protein function, sequence, pathways, genetic interaction, and nucleic acid. Biological data storage, analysis, and retrieval are known as bioinformatics. However, data mining is concerned with data analysis for discovering hidden relationships and trends in data.

State-of-the-art data mining techniques [2] can be categorized based on the mined knowledge type into decision trees, association rules, and clustering. However, biology databases have lacked the schemes to analyze massive information repositories, including genome databases [3]. Every data mining method has several algorithms, including (i) the association rule mining, using the Apriori or partition algorithms; (ii) the clustering techniques, using k-medoids or k-means algorithms; and (iii) the classification rule mining via decision tree generation using entropy value or gini index.

Progressive data mining techniques have powered post-genome biological research areas recently. Biological data mining offers inclusive data mining models, schemes, and applications in existing medical and biological research. The current chapter incorporates contributions from previous related work in the biological data mining research domain. It embraces the biological data mining concept and its applications. In addition, this work discusses the challenges and openings for analyzing and mining biological sequences/structures. The relationships between data mining and related computing techniques for biological data are also included. Furthermore, this chapter addresses data mining applications in bioinformatics domain.

The remaining sections of the current chapter are organized as follows. Section 12.2 reports on bioinformatics that uses data mining techniques. Section 12.3 includes various data mining techniques. Section 12.4 introduces the concept of data mining for biological sequences. Section 12.5 addresses some biological data mining applications. The challenges and future suggestions are introduced in Section 12.6. Finally, the conclusion is offered in Section 12.7.

## 12.2 Bioinformatics Using Data Mining Techniques

Bioinformatics is interested with the management, integration, mining, and interpretation of information from biological data. It uses computer technology to collect, store, recognize, extract, analyze, and combine biological data.

The development of data mining methods is an active research area in bioinformatics to solve biological data analysis problems [4]. There are several biological data analysis types, including protein structure prediction, cancer classification, gene classification, and protein structure prediction, which are based on gene expression data analysis clustering, microarray data analysis, and protein–protein interaction statistical modeling. The observations for labeling the regulatory elements and genes locations on each chromosome is essential in order to represent the datasets for the entire genomes of DNA sequence. Through bioinformatics, sequence analysis and genome annotation can be performed. Several bioinformatics techniques are incorporated for sequence analysis to define the biological function as well as the proteins code. In addition, genome annotation identifies the genes locations and the coding regions for understanding the species' genome.

Proteins have a vital role in all biological processes, with an extensive range of functions. Proteins must fold to function [5]. The protein's amino acid sequence can be obtained from the gene's sequence through the prediction of the protein structure.

Consequently, hidden predictive information extraction from massive databases using data mining to find relationships and patterns in the bioinformatics applications has a significant role. Commonly, knowledge/data discovery known as data mining process is employed for analyzing data from several perceptions and summarizations to convenient information. Figure 12.1 demonstrates the data mining steps, namely, (i) extract, transform, and load transaction data; (ii) store and manage the data in the

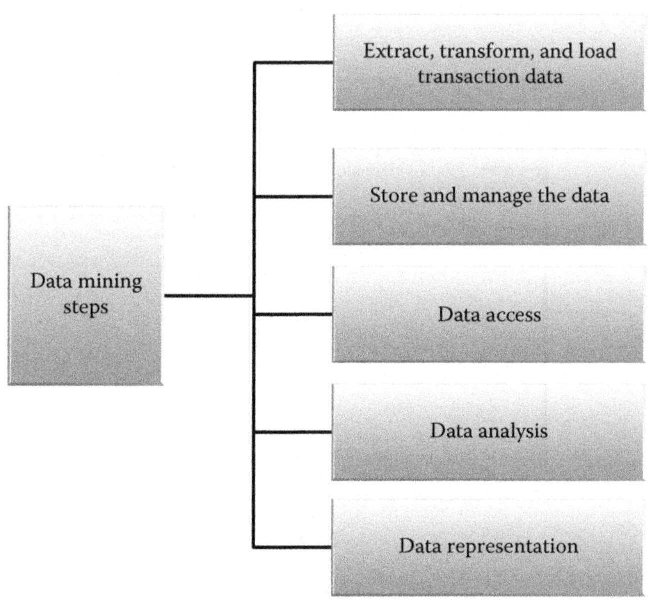


**FIGURE 12.1**
The steps of data mining process.

databases; (iii) provide data access; (iv) analyze the data; and (v) present the data in a meaningful and useful format.

Mining biological data is essential for extracting significant knowledge from the huge biology datasets. Data mining applications in bioinformatics embrace data cleansing, protein subcellular location prediction, protein function inference, protein function domain detection, disease prediction and treatment, gene finding, and gene/protein interaction network renovation. Accurate prediction is potentially a support to better patient treatment [6]. Various machine learning and data mining methods can be applied for peptide recognition.

## 12.3 Data Mining Techniques

The process of finding valuable patterns from massive data by converting data collection into knowledge is defined as data mining. The data mining concept can be divided into (i) predictive techniques including regression, classification, and time series analysis, and (ii) descriptive techniques to predict future prominence before occurrence, as illustrated in Figure 12.2.

As depicted in Figure 12.2, the predictive techniques are as follows:

- *Classification techniques*: Classification maps data into specific classes based on the data attribute values. Pattern recognition is considered to be a supervised learning classification type, as an input pattern is categorized into one class founded on its correspondence to the predefined classes [4–18]. Several methods can be used for the classification process, including decision trees, neural networks, support vector machines, and, Bayesian classifiers.

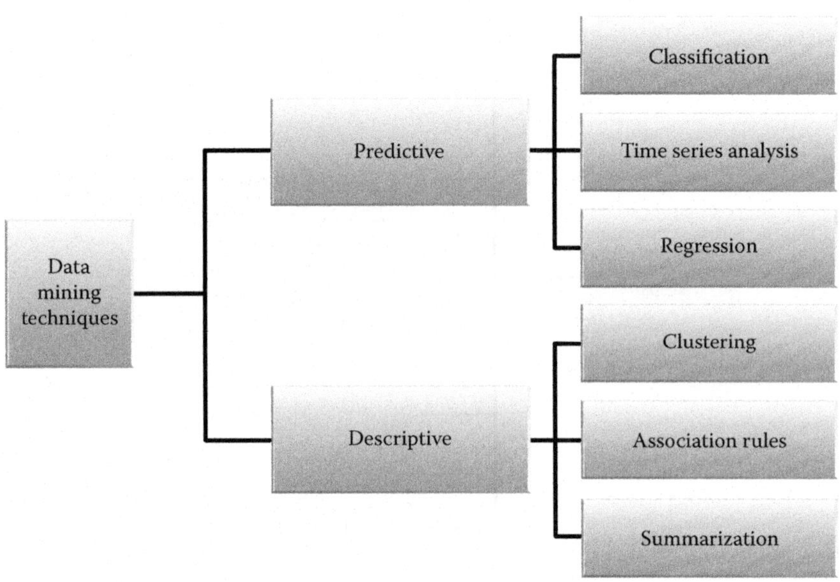


**FIGURE 12.2**
The data mining techniques.

- *Time series data analysis techniques*: Time series analysis considers temporal data objects that definitely attain from scientific applications. It includes univariate and multivariate time series techniques [19–24].
- *Regression methods*: Generally, regression is applied to expect forthcoming values founded on former values by fitting a values' set to a curve. It has a variety of methods [25–27] such as (i) nonparametric regression without any linearity assumption, (ii) robust regression analysis using a set of fitting criteria, (iii) ridge regression is the most generally used regularization technique of ill-posed problems, and (iv) nonlinear regression.

Moreover, the descriptive methods designate all the data using several methods, including the p-dimensional space partitioning into groups for segmentation and cluster analysis, the inclusive probability distribution models of the data for density approximation, and the descriptive models for the relationship between variables. Such techniques are as follows:

- *Clustering methods*: Clustering is analogous to classification, but it is defined by the data alone without predefined groups, which is considered an unsupervised learning. It segments (partitions) the data into groups that may be disjointed using the similarity amongst the data on predefined attributes [28–30]. Sophisticated clustering techniques include K-means clustering, and Fuzzy c-means clustering.
- *Association rules*: Association rule mining consists in searching for significant associations in a specified dataset using several methods such as the Apriori algorithm, quantitative association rules, multidimensional association rules, and distance-based association rules [31,32].
- *Summarization methods*: Summarization is used to find a description for a subset of data for automated report generation and interactive exploratory data analysis [33]. It has several methods including multivariate visualization methods, the derivation of summary rules, and the functional relationships discovery between variables.

## 12.4 Data Mining for Biological Sequences

The huge amount of gene data, nucleotides, microarrays, and peptide sequences of fungi, bacteria, virus, and other organisms produce valuable information about disease procedures. Data mining methods and machine learning techniques for information extraction from data are essential. Typically, classification is an initial step to examine a set of cases that which can be grouped based on their similarity. Data mining techniques/tools for biomolecular sequences and data classification such as WEKA, SVM (support vector machine), and Fuzzy-sets can be employed [34–37]. The intelligent bioinformatics community emphasizes on data-mining tools and systems in order to transform biological observations, sequences, and knowledge into structured information for visualization. Biological databases mining is challenging for mining biological sequences to develop computational framework for genomic DNA sequence and automatically yield a comprehensive annotation of the organism. It provides biologists with the ability of probing genomic data in detail with a broad range of viewpoints.

A relation exists between the DNA understanding process and the pattern recognition computational problems, machine learning, and information extraction using data mining. Researchers are interested in intelligent systems to solve leading computational–genomics problems, such as genome annotation to identify and classifying genes, computational comparative genomics to compare complete genomic sequences at different levels, and genomic patterns including regulatory regions identification in sequence data. These problems are essential to understand the biological organisms' function and their collaborations with the environment. The understanding of genes facilitates development of new treatments of genetic diseases, innovative antibiotics, and other drugs. Biological sequence mining is applied to discover a precise model of any organism's genome structure to provide informative characteristics for the sequence with its meaning.

Several classification procedures have been applied for protein sequence classification to specific classes, and subclasses, to extract features and to match these features' values in order to classify the protein sequence. Researches focus on developing various classification methods, such as the neural network model, and the rough set classifier.

Wu et al. [38] proposed a neural network (NN) model to classify protein sequences. N-gram encoding scheme of the input sequence was applied to extract features for constructing the pattern matrix that was used as input to the NN classifier. The results achieved 90% accuracy level. Zainuddin and Kumar [39] developed an advance method of [38]. The authors suggested initially using the 2-gram encoding technique to construct the pattern matrix. If this matrix was incapable of classifying the input protein sequence, then 3-gram encoding scheme results were added to the pattern matrix to be used for further matching using the NN.

Rao et al. [40] applied a probabilistic NN model using self-organized map (SOM) network to discover the relationships within protein sequences set through clustering them into several groups. The input to the NN first layer is the input sequences. Yellasiri and Rao [41] classified the massive protein data founded on functional and structural protein properties. A rough set protein classifier provided 97.7% accuracy. The authors proposed an innovative method called sequence arithmetic (SA) for identifying information and utilizing it to reduce the domain search space. Rahman et al. [42] used rough set classifier for extracting the necessary features for classification based on combining percentages of 20 amino acids properties. For data mining and knowledge discovery, the Rosetta system was used.

## 12.5 Biological Data Mining Applications

Data mining applications to bio-data mining contains protein function domain detection, gene discovery, function motif recognition, protein function interpretation, data cleansing, disease treatment optimization, disease diagnosis and prognosis, gene/protein interaction network restoration, and protein subcellular position estimation. The protein constructing blocks are the amino acids. Twenty different amino acids are applied to produce proteins. Each protein shape and other characteristics are dictated by the specific amino acids sequence in it. The amino acids chemical properties of proteins control the protein biological activity.

For massive genomic data analysis, Anandhavalli et al. [43] were interpreted with explaining how the expression of any specific gene might influence other genes' expression. Huang et al. [44] identified genes' expression using lung adenocarcinoma tumor and neighboring nontumor tissues dataset. The authors applied microarray data analysis, protein–protein collaboration network, and cluster analysis.

Win et al. [45] examined thousands of genes concurrently for microarray gene expression and survivability forecast. Dimensionality decreasing was considered for discretization and selection process. This approach delivered more system generalization fitness and less computational complexity. This study was insufficient in calculating lung cancer progress and clinical outcome.

Deoskar et al. [46] suggested the SPACO (support-based ant colony optimization) method for lung cancer symptoms. Significant patterns were extracted, and then frequent symptoms were selected by support count value. The authors established that either decreasing or increasing the patient prediction level helped in distinguishing lung cancer and improved the accuracy. Shukla et al. [47] studied various data mining methods, including classification, clustering, association rule mining, and regression, which are extensively applied in healthcare domain. These methods are used to develop the prediction and diagnosis qualities for different diseases using several algorithms such as the genetic algorithm, k-means clustering, association rule mining, and naïve Bayesian classification, applied to massive medical data volume.

Mao and Mao [48] explored the A Priori-Gen procedure to study the disease association that had a large dataset to discover the association among multiple single nucleotide polymorphisms (SNPs). The disease association study aims to assess collected information to discover the multi-SNPs' interaction associated to compound diseases with statistical power and high accuracy. Martinez et al. [49] proposed the GENMINER method to extract the association rules from genomic data. The authors used the NORDI algorithm for minimal nonredundant rules extraction.

Cancer is the most significant cause of death. Lung cancer occurrence is increasing rapidly. Bio-data mining is the procedure of extracting implicit, nontrivial, formerly unknown, and theoretically valuable patterns/information from large biological sequences amount. There are numerous data mining systems for association rule mining currently used in the biological science domain. In the biological datasets, in order to discover the association among the sets' items, data mining can be used. Kalaiyarasi and Prabasri [50] predicted the dominant amino acids which cause the lung cancer. In association rule mining, numerous procedures are accessible for calculating frequent patterns. Nevertheless, a few procedures have definite drawbacks, for instance, space/time complexity and high cost. These drawbacks could be rectified using developed data mining method that offers promising ways for fighting lung cancer.

Gene expression data can be employed to predict clinical results. Haferlach et al. [51] proposed a gene expression summarizing classifier to categorize patients into 18 diverse subclasses of either lymphoid or myeloid leukemia. The all-pairwise classification system was proposed using the trimmed mean of the difference between mismatch and perfect match intensities with quantile normalization. The difference of quantile normalized values (DQN) technique was clarified in [52,53]. Salazar et al. [54] constructed a gene expression classifier and extracted the gene features using leave-one-out cross-validation technique to define which gene probes were powerfully correlated with the metastasis-free survival (DMFS) with a t-test as the conclusive factor.

## 12.6 Evolution, Challenges, and Future Scope

Both bioinformatics and data mining are fast-intensifying and closely related research areas. It is imperative to inspect the significant research topics in bioinformatics and to improve innovative data mining techniques for effective and scalable biological analysis. Given the problems of biological data mining and analysis, bioinformatics scientists can consider the following computational difficulties for future study:

- Improving sequence-pattern discovery algorithms.
- Evolving new approaches of bootstrapping learning algorithms from the biological data.
- Developing machine learning algorithms for outsized sequence sources.
- Incorporating multiple information sources into an integrated learning and data mining system.
- Improving the accuracy and speed of the probabilistic-reasoning systems.
- Including optimization algorithms such as the genetic algorithm, particle swarm optimization, and cuckoo search algorithm for enhanced data mining systems. For example, genetic algorithms can be applied to the association and classification methods.
- Techniques can be employed to discover associations among similar gene clusters, genes, protein sequences, and using decision trees for genes classification.
- Evolving approaches for intelligent selection of the accurate states set from the numerous Markov models is an open research area.
- For biological sequences analysis and processing, in addition to sequence relations, efficient classifiers must be considered. The information account with relative position of the different shared features should be considered. One of the future objectives is to improve features that can exploit position-precise information.

In the biomedical domain, massive datasets are accessible. Numerous algorithms for finding common patterns from the biological sequences are used to predict cancer. Some models use efficient frequent pattern procedure to mine the most recurrent patterns from the specified input dataset to find the most controlling amino acid sequence in order to block the cancer cells growing from the clustered protein sequence. In conclusion, the expected amino acids could be more valuable in making medicine for curing lung cancer. Consequently, existing cancer research is investigating several protein sequences, including tyrosine kinase, ALK, Ral protein, and histone deacetylase sequence, which can be used to block the cancer cells' growth.

## 12.7 Conclusion

Data mining methods are applied to discover significant formulations and correlations from formerly collected data. Numerous diverse application domains exploit data mining as a means for achieving actual usage of the internal information. Data mining becomes increasingly more extensive in the public and private sectors, such as in industries, insurance, banking, retailing, and medicine, to enhance research and reduce costs. A variety of methods such as time series analysis, decision tree models, and regression were in use before the term data mining became common. Nevertheless, there are also methods that were found by data mining specialists in the last decade, including SVM, Apriori algorithm and, c-means clustering.

Numerous application domains of extrapolative approaches are associated with medicine fields. They have become progressively prevalent with the growth of biotechnology in the last decade. Most of the genetics research is directed toward data mining technology. Descriptive methods are commonly used in banking, finance, and social sciences to designate a definite population. Clustering is the most common description technique. In the last decade, the k-means technique has lost popularity as compared to the c-means procedure. An additional common technique is the association rules, where Apriori is the most desired technique.

The association rules are still have important role due to the increased databases and the information production resources. Sequence discovery is also a rising domain, currently. Various challenges and future perspectives were introduced in this chapter to guide researchers for the further development of the most critical topics in the field.

## References

1. Vasantha Kokilam, K. and Pon Mary Pushpa Latha, D. (2012), A review on evolution of data mining techniques for protein sequence causing genetic disorder diseases, *2012 IEEE International Conference on Computational Intelligence & Computing Research (ICCIC)*, pp. 1–6, IEEE.
2. Pujari, A. (2001), *Data Mining Techniques*. Nancy, France: Universities Press.
3. Zhang, D. and Zhou, L. (November 2004), Data mining techniques in financial application, *IEEE Transactions on Systems, Man and Cybernetics—Part C: Applications and Reviews*, 34(4), 513–522.
4. Chen, J.Y., Zaki, M.J., and Lonardi, S. (2008), BIOKDD08: A workshop report on data mining in bioinformatics, *SIGKDD Explorations*, 10(2), 54–56.
5. Richard, R.J.A. and Sriraam, N. (2005), A feasibility study of challenges and opportunities in computational biology: A Malaysian perspective, *American Journal of Applied Sciences*, 2(9), 1296–1300.
6. Lee, K. (2008), Computational study for protein-protein docking using global optimization and empirical potentials, *International Journal of Molecular Sciences*, 9, 65–77.
7. Kriti, J.V., Dey, N., and Kumar, V. (2015), PCA-PNN and PCA-SVM based CAD systems for breast density classification. In: Hassanien, A.-E. Grosan, C. and Tolba, M.F. eds. *Applications of Intelligent Optimization in Biology and Medicine: Current Trends and Open Problems*, Springer International Publishing. Springer, Berlin, 96, 159–180.
8. Kausar, N., Palaniappan, S., Al Ghamdi, B.S., Samir, B.B., Dey, N., and Abdullah, A. (2015), Systematic analysis of applied data mining based optimization algorithms in clinical attribute extraction and classification for diagnosis of cardiac patients In: Hassanien, A.-E. Grosan, C. and Tolba, M.F. eds. *Applications of Intelligent Optimization in Biology and Medicine: Current Trends and Open Problems*, Springer International Publishing. Springer, Dordrecht, 96, 159–180.

9. Dey, N. and Ashour, A. eds. (2016), *Classification and Clustering in Biomedical Signal Processing*, Advances in Bioinformatics and Biomedical Engineering (ABBE) Book Series IGI Global.
10. Saba, L., Dey, N., Ashour, A.S., Samanta, S., Nath, S.S., Chakraborty, S., Sanches, J., Kumar, D., Marinho, R.T., and Suri, J.S. (2016), Automated stratification of liver disease in ultrasound: An online accurate feature classification paradigm, *Computer Methods and Programs in Biomedicine*, 130, 118–234.
11. Ahmed, S.S., Dey, N., Ashour, A.S., Sifaki-Pistolla, D., Bălas-Timar, D., and Balas, V.E. (2016), Effect of fuzzy partitioning in Crohn's disease classification: A neuro-fuzzy based approach, *Medical & Biological Engineering & Computing*, 55(1), 101–115.
12. Ghosh, A., Sarkar, A., Ashour, A.S., Balas-Timar, D., Dey, N., and Balas, V.E. (2015), Grid color moment features in glaucoma classification, *International Journal of Advanced Computer Science and Applications (IJACSA)*, 6(9), 1–4.
13. Nath, S., Kar, J., Chakraborty, S., Mishra, G., and Dey, N. (July 2014), A survey of image classification methods and techniques, *International Conference on Control, Instrumentation, Communication and Computational Technologies-2014*, pp. 10–11.
14. Dunham, M. (2003), *Data Mining: Introductory and Advanced Topics*, Upper Saddle River, NJ: Prentice Hall.
15. Armand, S., Watelain, E., Mercier, M., Lensel, G., and Lepoutre, F.X. (2006), Identification and classification of toe-walkers based on ankle kinematics, using a data-mining method, *Gait & Posture*, 23, 240–248.
16. Lee, T.S., Chiu, C.C., Chou, Y.C., and Lu, C.J. (2006), Mining the customer credit using classification and regression tree and multivariate adaptive regression splines, *Computational Statistics & Data Analysis*, 50, 1113–1130.
17. Nitanda, N., Haseyama, M., and Kitajima, H. (2004), An audio signal segmentation and classification using fuzzy c-means clustering, *Proceedings of the Second International Conference on Information Technology for Application*.
18. Pan, F., Wang, B., Hu, X., and Perrizo, W. (2004), Comprehensive vertical sample-based KNN/LSVM classification for gene expression analysis, *Journal of Biomedical Informatics*, 37, 240–248.
19. Swift, S. and Liu, X. (2002), Predicting glaucomatous visual field deterioration through short multivariate time series modeling, *Artificial Intelligence in Medicine*, 24, 5–24.
20. Chen, O., Zhao, P., Massaro, D., Clerch, L.B., Almon, R.R., DuBois, D.C., Jusko, W.J., and Hoffman, E.P. (2004), The PEPR GeneChip data warehouse, and implementation of a dynamic time series query tool (SGQT) with graphical interface, *Nucleic Acids Research*, 32, 578–581.
21. Cuaresma, J.C., Hlouskova, J., Kossmeier, S., and Obersteiner, M. (2004), Forecasting electricity spot-prices using linear univariate time-series models, *Applied Energy*, 77, 87–106.
22. Kim, S., Imoto, S., and Miyano, S. (2004), Dynamic Bayesian network and nonparametric regression for nonlinear modeling of gene networks from time series gene expression data, *Biosystems*, 75, 57–65.
23. Liao, T.W. (2003), Clustering of time series data—A survey, *Pattern Recognition*, 38, 1857–1874.
24. Romilly, P. (2005), Time series modelling of global mean temperature for managerial decision-making, *Journal of Environmental Management*, 76, 61–70.
25. Mohanty, M., Painuli, D.K., Misra, A.K., Bandyopadhyaya, K.K., and Ghosh, P.K. (2006), Estimating impact of puddling, tillage and residue management on wheat (*Triticum aestivum*, L.) seedling emergence and growth in a rice–wheat system using nonlinear regression models, *Soil and Tillage Research*, 87, 119–130.
26. Roberts, S. and Martin, M. (2005), A critical assessment of shrinkage-based regression approaches for estimating the adverse health effects of multiple air pollutants, *Atmospheric Environment*, 39, 6223–6230.
27. Zenkevich, I.G. and Kránicz, B. (2003), Choice of nonlinear regression functions for various physicochemical constants within series of homologues, *Chemometrics and Intelligent Laboratory Systems*, 67, 51–57.
28. Chen, M.C. and Wu, H.P. (2005), An association-based clustering approach to order batching considering customer demand patterns, *Omega*, 33, 333–343.

29. Oatley, G.C. and Ewart, B.W. (2003), Crimes analysis software: "pins in maps," clustering and Bayes net prediction, *Expert Systems with Applications*, 25, 569–588.
30. Sebzalli, Y.M. and Wang, X.Z. (2001), Knowledge discovery from process operational data using PCA and fuzzy clustering, *Engineering Applications of Artificial Intelligence*, 14, 607–616.
31. Delgado, M., Sánchez, D., Martín-Bautista, M.J., and Vila, M.A. (2001), Mining association rules with improved semantics in medical databases, *Artificial Intelligence in Medicine*, 21, 241–245.
32. Zhang, S., Lu, J., and Zhang, C. (2004), A fuzzy logic based method to acquire user threshold of minimum-support for mining association rules, *Information Sciences*, 164, 1–16.
33. Kantardzic, M. (2011), *Data Mining: Concepts, Models, Methods, and Algorithms*, John Wiley & Sons.
34. Liu, Y. and Wan, X. (2016), Information bottleneck based incremental fuzzy clustering for large biomedical data, *Journal of Biomedical Informatics*, 62, 48–58.
35. Villalba, S.D. and Cunningham, P. (2007), An evaluation of dimension reduction techniques for one-class classification, *Artificial Intelligence Review*, 27(4), 273–294.
36. Rajapakse, J.C. and Ho, L.S. (2005), Markov encoding for detecting signals in genomic sequences, *IEEE/ACM Transactions on Computational Biology and Bioinformatics*, 2(2), 131–142.
37. Ashoor, H., Mora, A.M., Awara, K., Jankovic, B.R., Chowdhary, R., Archer, J.A.C., and Bajic, V.B. (2011), Recognition of translation initiation sites in *Arabidopsis thaliana. Systemic Approaches in Bioinformatics and Computational Systems Biology*, Recent Advances: Recent Advances. pp. 105–116.
38. Wu, C., Berry, M., Shivakumar, S., and Mclarty, J. (1995), *Neural networks for full-scale protein sequence classification: Sequence encoding with singular value decomposition*, 21(1-2), 177–193.
39. Zainuddin, Z. and Kumar, M. (2008), Radial basic function neural networks in protein sequence classification, *Malaysian Journal of Mathematical Science*, 2(2), 195–204.
40. Nageswara Rao, P.V., Uma Devi, T., Kaladhar, D., Sridhar, G., and Rao, A.A. (2009), A probabilistic neural network approach for protein super-family classification, *Journal of Theoretical and Applied Information Technology*, 6(1), 101–105.
41. Yellasiri, R. and Rao, C.R. (2009), Rough set protein classifier, *Journal of Theoretical and Applied Information Technology*.
42. Rahman, S.A., Bakar, A.A., and Hussein, Z.A.M. (2009), Feature selection and classification of protein subfamilies using rough sets, *International Conference on Electrical Engineering and Informatics*, Selangor, Malaysia.
43. Anandhavalli, M., Ghose, M.K., and Gauthaman, K. (2010), Association rule mining in genomics, *International Journal of Computer Theory and Engineering*, 2(2), 269.
44. Huang, C.-H., Wu, M.-Y., Chang, P.M.-H., Huang, C.-Y., and Ng, K.-L. (2014), In silico identification of potential targets and drugs for non-small cell lung cancer, *IET Systems Biology*, 8(2).
45. Win, S.L., Htike, Z.Z., Yusof, F., and Noorbatcha, I.A. (June 2014), Gene expression mining for survivability of patients in early stages of lung cancer, *International Journal of Bioinformatics and Biosciences*, 4(2).
46. Deoskar, P., Singh, D., and Singh, A. (September 2013), An efficient support based ant colony optimization technique for lung cancer data, *International Journal of Advanced Research in Computer and Communication Engineering*, 2(9).
47. Shukla, D.P., Patel, S.B., and Sen, A.K. (February 2014), A literature review in health informatics using data mining techniques, *International Journal of Software and Hardware Research in Engineering*, 2, 2347–4890.
48. Mao, W. and Mao, J. (2009), The application of Apriori-Gen algorithm in the association study in type 2 diabetes. In: *3rd International Conference on Bioinformatics and Biomedical Engineering*, (ICBBE) 2009, pp. 1–4. IEEE.
49. Martinez, R., Pasquier, C., and Pasquier, N. (2010), GENMINER: Mining informative association rules from geenomic data, *IEEE International Conference on Bioinformatics and Biomedicine*.
50. Kalaiyarasi, R. and Prabasri, S. (2015), Predicting the lung cancer from biological sequences, *International Journal of Innovations in Engineering and Technology*, 5(1).

51. Haferlach, T., Kohlmann, A., Wieczorek, L., Basso, G., Kronnie, G.T., Béné, M.C., De Vos, J. et al. (2010), Clinical utility of microarray-based gene expression profiling in the diagnosis and subclassification of leukemia: Report from the international microarray innovations in leukemia study group, *Journal of Clinical Oncology*, 28(15), 2529–2537.
52. Liu, W., Li, R., Sun, J.Z., Wang, J., Tsai, J., Wen, W., Kohlmann, A., and Williams P.M. (2006), PQN and DQN: Algorithms for expression microarrays, *Journal of Theoretical Biology*, 243(2), 273–278.
53. Bennett, K.P. and Campbell, C. (2000), Support vector machines: Hype or hallelujah, *SIGKDD Explorations Newsletters*, 2(2), 1–13.
54. Salazar, R., Roepman, P., Capella, G., Moreno, V., Simon, I., Dreezen, C., Lopez-Doriga, A. et al. (2011), Gene expression signature to improve prognosis prediction of stage II and III colorectal cancer, *Journal of Clinical Oncology*, 29, 17–24.

# 13

## *Video Text Extraction and Mining*

**Surekha Borra, Nilanjan Dey, and Amira S. Ashour**

**CONTENTS**



**ABSTRACT** With the rapid growth in computer communication, storage, and networking technology, discovering and extracting interesting features and patterns for video classification and mining is on the rise. Text in video sequences provides complementary but imperative information for video retrieval and indexing. This chapter aims at the discussion of the extraction of text information from video and multi-modal mining from the same. This chapter classifies and briefs the methods used to extract text from videos, discusses their performance, mentions their merits and drawbacks, enlists available databases, their vulnerabilities, and challenges, and provides recommendations for future development.

**KEY WORDS:** *video processing, video mining, multimedia, text extraction.*

## 13.1 Introduction

Video is a type of multimedia data, and is a combination of audio, visual, text, and metadata. Video is composed of a sequence of images varying temporally. Videos play an imperative role in sports, entertainment, education enhancement, medical applications, and surveillance. The accumulation of countless videos, both online and off-line, has triggered the research community to focus on development of video mining and automatic analysis of videos by text extraction and recognition approaches.

Video mining [1] deals with identifying, extracting, and describing the hidden knowledge, interesting points, features, and visual patterns from an extremely large database of videos. Lack of a priori knowledge about the occurrence of visual patterns, temporal uncertainties, and their complex structural properties makes video mining a challenging task when compared to other statistical and text-based techniques. Hence, the development of video mining tools has become crucial in many applications involving decision-making.

Detection and recognition of mechanically added text over video frames help in automatic annotation, indexing, and images structuring. Text extraction from video differs from object recognition in methodology and complexity. Traditional optical recognition technology (OCR) cannot be directly used to extract text from videos, as the accuracy depends on the localization of text in frames. While most researchers use state-of-the-art document image processing algorithms, or object detection and tracking methods for this purpose, very few methods exploit temporal information. While most of the work is done on static caption/scene text extraction, very few methods are focused on scrolling text.

Video text [2] is usually of three types: caption text, scene text, and embedded text. Scene text is the naturally existing text in the scene captured by the video recorder. Examples include banners, buildings, shirts, and signs. Embedded text is embedded and overlaid onto the frame. The caption text is artificial text and can be static or moving. It is usually superimposed on the video frames after recording the scenes. In general, in news video, a caption text directly describes the content presented/person's names, subtitles, and language translation text. These are mostly superimposed on a fixed position of the screen. Other names for caption text are overlay text, mechanical text, and graphic text. The extraction of this kind of text is mainly useful in creating indices.

Text extraction from video is extremely difficult [3] as it involves several assumptions about the input to the system:

- Single frame/sequence of frames
- Gray/color frame
- Compressed/uncompressed frame
- Static/moving text
- Text font size, style, alignment, color, orientation of fonts, background

Video caption text extraction methods are generally classified into two groups namely, traditional approaches and training approaches:

1. Traditional techniques make use of a single or a combination of features extracted from image/video. The features can be local or global. Examples include edges, intensity, shape, texture, color, shape, and temporal changes.
2. Training techniques can automatically extract text by making use of well-defined classifiers and machine learning algorithms. The support vector machine (SVM)–based systems have shown better performance in real-time applications as they are based on statistical learning theory. The performance of these algorithms is better in cases where there are dynamic changes in text size, font, layout, texture, color, etc.

Further, there are some advanced learning methods such as neural networks–based approaches and hybrid methods which benefit from both traditional and training methods.

Basic background in video mining is presented in Section 13.2. Traditional and modern training methods are addressed in Section 13.3. Case studies are covered in Section 13.4, followed by performance measures in Section 13.5. The challenges and issues are collectively described in Section 13.6, and future directions are listed in Section 13.7.

## 13.2 Video Mining

Video data mining aims at automatic extraction from large video databases, the implicit knowledge, the structure and contents of videos, static or moving objects, their characteristics, spatial and temporal relations among the objects, and activities and events without prior knowledge. It is different from video understanding where the aim is to process videos for extraction of features from a video database. Video understanding is performed by selecting features from multiple modalities and bridging the gap between them. In general, there are three kinds of modalities: visual, audio, and text, which deal with the scene in the video, speech, sounds of video and textual resources that describe the video content. A variety of low-level, semantic, and syntactic information and features can be extracted from videos that describe the shape, texture, and color of the static and moving objects in the scene, their spatial–temporal positions and relations, their meaning, and actions.

Video processing for mining involves conversion of a nonstructured video into structured data using audio processing, image processing, pattern recognition, computer vision, artificial intelligence, and neural network techniques. The video data can be structured [4] in two ways, according to the content structure.

1. Scripted video databases are generated with a plan of editing and compiling them later for storage and distribution. Examples include news videos, dramas, and movies.
2. Unscripted videos are usually live videos that do not follow any plan. Examples include sports and surveillance videos.

Advantages of using video mining for video retrieval are as follows:

- Effective in searching media content
- Saves storage and computational complexity due to dimensionality reduction
- Design of algorithms that fill the gap between low/medium and high-level features
- Allows multiple modalities for indexing video

There are three main steps in video mining:

1. *Video segmentation and preprocessing*: It involves the selection of significant shots/scenes/frames/objects. A video track is partitioned into key shots, and selecting a better frame of each shot for further image processing. It enables a

preprocessing stage that involves noise removal and enhancement. The video segmentation step also allows identification of shot boundaries, zooms, and pans. Several data models [5] such as the content-based retrieval model and hierarchical model are presented in the literature for video representation based on their contents and applications.

2. *Feature extraction and description*: Once the video is segmented for key frames, each key frame is treated as an image, and processes for the extraction of semantic content-based features, which can be local or global. Examples include edges, textures, shapes, colors, color histograms, color correlograms, positions, and sizes of objects in the key frame. At a high level, the content-based features can make use of object classes and their trajectories. Another kind of feature is a description-based feature that employs meta-data such as captions, keywords, video creation time, and video size. In addition, multimodal features such as audio keywords and spatial—temporal features can be combined to analyze the patterns and to extract the knowledge of the application.
3. Discovering knowledge and patterns by combining multimodal features extracted from audio, video, and text.

### 13.2.1 Applications of Video Mining

*Video retrieval* [6]: It is a kind of database management system where the input is a video query and the outputs are similar video clips. The steps involved are video segmentation for key frame selection, low-level feature extraction from each frame, dimensionality reduction, and the content-based video retrieval from the database of videos. The performance of these systems depends on the manipulation of video data.

*Video abstraction* [7]: It is a means of creating a short video summary. Two widely used approaches for video abstraction are video summarization and video skimming. The summary can be in terms of a sequence of significant frames or video skims that contain maximum information. A video abstract allows effective navigation in minimum time.

*Video annotation* [8]: This involves manually or automatically generating captions (or keywords) and embedding them into video frames with an objective of filling the semantic gap between multiple modes (text, images, and video). It is a powerful tool as it allows video retrieval from text query. Two widely used approaches are supervised and unsupervised learning. Annotations are also categorized based on context, statistics, and rules [9].

### 13.2.2 Challenges in Video Mining

Video mining is complex [7,10] mainly due to the unstructured nature of videos. Some of the research directions in video mining are listed here:

- *Video semantic event detection*: The problem of predicting semantic features while resolving the semantic gap is an important research problem, where it finds its application in automatic detection of suspicious events in surveillance videos, content-based advertisements and content filters.

- Development of a general framework irrespective of type of video and domain (sports, news, etc.) that can extract multiple semantics is needed.
- Modeling multimodal characteristics for event detection is useful in applications in medical, traffic, and sports videos where temporal patterns, motion trajectory, and tracking plays a vital role.
- Discovering hybrid features to improvise the results of video data mining.
- Discovering a universal segmentation technique irrespective of the type of visual media.
- Development of algorithms for recognizing and tracking significant objects/ scenes.
- Discovery of tools for finding correlation among significant objects/scenes of the video.
- Development of video database models for effective mining and management.
- Resolving the semantic gap between multimodal features and user domain.
- Development of optimized techniques for video mining and retrieval that can perform effective semantic interpretation of the video.
- Identifying hybrid data mining approaches that can deal with unstructured videos directly.

## 13.3 Video Text Extraction

Video texts obtained from speech transcriptions, embedded text, caption text, and scene text represent most of the semantic information and can reveal the significant events and actions in a video.

### 13.3.1 Applications of Video Text Extraction

With the invention of smart phones, the applications of video text extraction are on the rise. Video understanding, retrieval, annotation, and indexing are helpful in browsing the internet for interested videos. Apart from multimedia retrieval, text extraction from videos has several applications which are listed here:

1. *Elimination of moving text on the screen* [11]: Elimination/masking of moving text lines from offline/online videos is required for viewers' convenience as the information related to broadcast programs, channels, and advertisements is usually displayed on the video frames during broadcasting and are annoying in most cases.
2. *Language translation* [12]: Apps that are capable of recognizing the text and translating them in real time into the arterial language from captured videos and images are in great demand now with the progress of smart phone technology. Multimedia users who use personal digital assistants can translate images having foreign text or signs in to their own language. Such applications are a good aid to travelers visiting foreign countries.

3. *Text-to-speech conversion* [13]: Developing personal text-to-speech devices and apps assists visually impaired and blind people. Such kind of devices also assists in understanding important information such as instructions, warning signs and labels.
4. *Automation* [14]: Text extraction and recognition improves efficiency and reduces manual labor by automating the sorting of addresses, houses, containers, packages, etc. Automatic detection and text extraction in maps is widely applied in geo-coding systems.
5. *Television commercials* [15]: Automatic text extraction from video in real time eases the monitoring of the broadcasted advertisements for the count, date, and time of broadcast, in a specified channel. Some parents would NOT like their children to watch and get influenced by TV commercials. Using video mining techniques, it is possible to detect and stop TV commercials from playing.

Other possible applications are also listed:

- *Event detection in sports and games*: Real-time analysis of caption text in broadcasted sports videos is essential in producing sports video summarization, retrieval, and indexing.
- *Guiding robots* [16]: Robots can be guided based on the information extracted from the scene texts of surrounding views that are received from the on-board video camera.
- *Assisting visually impaired* [17]: A wearable camera, which is able to extract and convert text from scene into brail or speech in real time, is a most anticipated device for the visually impaired. Development of text extraction and tracking methods that are insensitive to fast camera movements is challenging.
- *Text -to-text translators/reality translator* [18]: Text written in an unknown language can now be converted into a required language using Smartphone apps. Improving the robustness and accuracy, and reducing the run time are the focus of research in language translators.
- *Real-time positioning system* [19]: Real-time tracking of a person is important for event detection and context awareness and for creating self-summaries. Wearable computers with dynamic video text extraction programs are capable of performing this task.
- *Real-time traffic monitoring* [20]: Real-time automatic extraction of license plate characters using adaptive threshold methods helps in managing traffic.
- *Assisting drivers* [21]: Real-time automatic detection of road signs and the related text assists drivers in providing information about permits, prohibits, risks and warnings, traffic density, and navigation.

### 13.3.2 Traditional Approaches

Most of the traditional approaches to caption text extraction are similar to text extraction from documents/scanned images, where the main focus is to reduce noise effects, improve text detection ability, segmentation, and OCR. The latest trends use hybrid approaches that combine and employ the advancements of computer vision, machine learning, and neural network technologies. Several optimization techniques [22–24] are also integrated and

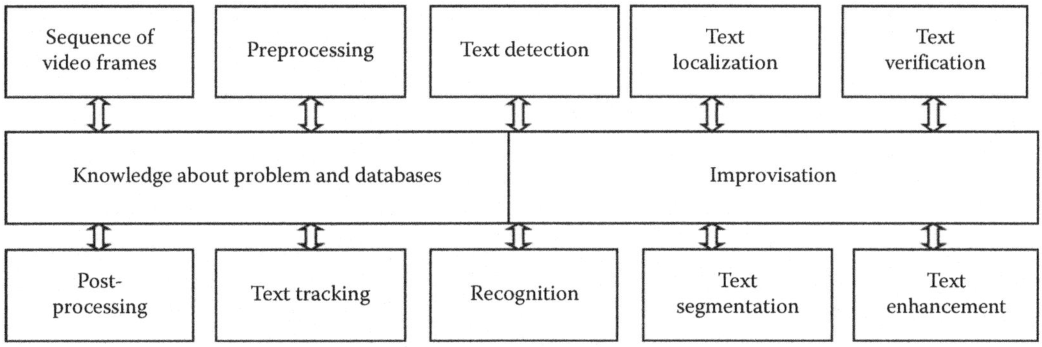


**FIGURE 13.1**
Steps in video text extraction.

implemented to design high-performance text extraction techniques to face complex situations.

Traditional video text extraction methods treat video frames independently for text detection and recognition, without considering temporal information between frames. These are particularly useful for extracting the scene text or embedded caption text where it is often completely confined to a single frame.

Extraction of text from video generally involves several steps [25] that are application dependent and are shown in Figure 13.1.

- *Preprocessing*: The first step in any approach after preprocessing is to detect if there is text in a given frame.
- *Text detection and localization*: This step detects and localizes the text in every frame with rectangular and color-bounding boxes. The localization divides the frame into region of interest (ROI), which is usually the text segments, and nonregion of interest (NROI) components. Localization detects the location in the frame, where text exists. Common features used at this stage are edges [26], color [27], and texture [28]. The recent trend is to use regions [29], points [30], and character appearance features [31]. After feature extraction, connected components based on pixel layout or color similarity, statistical analysis, and pattern recognition methods define and analyze the regions of text. The simplest way to overcome this problem is to convert the color image into intensity components before processing. Other ways are to work in other color spaces such as hue-saturation-value (HSV) [32], hue-lightness-saturation (HLS) [33], or with color layers generated from mean shift algorithm [34]. Another approach to locating text is to use Gaussian mixture models in intensity, hue and RGB channels [35]. Since text shows strong gradients, edge is well used in text localization. Features such as Gaussian derivatives and gradient vector flows are widely used in extracting vertical edges of text. Though the edges are less sensitive to brightness and color variations, they show poor performance for frames having sharp edges in the scene. A recent trend is to use a sliding window approach along with modern classifiers such as Adaboost [36]. A multiscale sliding window is also defined to locate characters that are closely spaced. Features extracted from a histogram of gradients (HOG), and from transform domains such as discrete cosine transform DCT [37], Fourier transform (FT) [38], and wavelet transform [39] are generally used as texture features. Another interesting feature

that is used in text extraction from video is Harris corners [30] which are corner point features of text. Constant color regions can be easily detected by the maximally stable extremal region (MSER), and pruning algorithms [40] are effective as text color strongly varies with background color. Hybrid features are often used to improve the performance [41]. Detection methods that are based on tracking consider temporal redundancy in addition to spatial information for noise removal and to develop the detection performance. A text trajectory is defined with length equal to the duration of the start and end frames of single text and is compared with a threshold to decide if the text being tracked is noisy or effective. Additional progress in the discovery accuracy can be realized with hybrid methods that combine the results of tracking and detection stages, along with multiple frame integration.

- *Text verification*: The ROI is then divided into text and nontext regions (e.g., spaces) through verification. This step is required to improve the performance of text detection. A variety of holistic features such as aspect ratio, edge density, distance between characters, LBP, HOG, wavelets, stroke widths, Gabor filters, and multilayer perceptrons are used to differentiate between text and nontext. Thresholding is another technique that is applied to distinguish the text and nontext and to reduce the false positive rate of text detection. Thresholds are applied on a variety of features such as vertical and/or horizontal projections [42], edges [43], aspect ratios [44], text block height, and distance between characters. Though the thresholding task is simple, it requires prior knowledge about the text to be extracted. Recently, deep learning techniques such as convolution neural networks (CNNs) are used in improving the performance of text verification [45].
- *Text extraction/segmentation*: This technique uses binarization to segment the characters/words and constructs independent blocks of characters/words to follow their recognition. The objective of binarization is to separate text pixels from the background. Adaptive thresholding [46] depends on the local features and is compatible to different backgrounds. To offer better performance Gaussian mixture models [47], multi-stage CRF models [48] were defined. Classification and clustering techniques suits best in case of low-quality videos. Gabor filters, K-means clustering techniques, graph cut algorithms, and skeleton cut algorithms showed best performance.
- *Segmentation*: Projection analysis is usually used to segment a detected text region into individual characters. Finding the optimal threshold for the projection is difficult when handling low-quality frames. While low thresholds lead to too much segmentation, high thresholds lead to missing segmentations. An optimization method [49] is to use features based on gradient vector flow and a two-path search algorithm. While searching in one direction is effective in reducing false alarms, searching in the other direction leads to true segmentations.
- *Text enhancement*: This method recovers distorted text and provides high-quality text.
- *Text recognition*: It recognizes the segmented text with optical character recognition (OCR) techniques. Obviously, the performance of recognition is based on the results of detection, binarization, and segmentation step. Employing video text recognition [50,51] with commercial OCR engines is not always a good choice as this can lead to false positives with video frames of low resolution, low contrast,

and complex backgrounds. One solution is to employ temporal redundancy of text tracking and integrate this with enhancement techniques. The objective is either to choose the high-resolution text region from the frame, or to improve the resolution of the text by image fusion techniques. While the former methods pose problems for blurred frames, the latter methods that employ multiframe averaging pose problems for blurred edges.

- *Tracking*: The objective is to continuously locate the position of scrolling text over a video frames sequence. This step integrates the recognized text that is scrolling across a sequence of video frames. Obviously, the text extraction performance is based on the detection and recognition steps results. Hence, tracking methods are broadly classified into two types [3]:
    - *Recognition-based tracking* (*RT*), wherein the recognized text from sequential frames are used for integrating the scrolling text. Features such as distance between the recognized words in the consequent frames are used as for text integration and matching. Extraction and analysis of temporal information along with spatial information from a frame sequence and integration of the text in a video improves the accuracy of text extraction.
    - *Detection-based tracking* (*DT*), wherein the locations of detecting bounding boxes from sequential frames are used for integrating the scrolling text. General methods used in this category are particle filtering, template matching, and tracking by detection.
    - *Tracking with template matching*: This method checks for some fixed unique text features (called templates) in each frame and their locations. The template can be a patch, edge, or even a cumulative histogram of intensity. Text tracking methods based on template matching show good performance with low-resolution frames, scrolling texts, and complex backgrounds. Feature selection plays an important role in improving the performance of tracking. While the color features survive against multioriented text, multiscale text, and image deformations, they are not robust with color bleeds due to compression or illumination changes. Color features exhibit poor performance when there are similar text colors in the background. The edge features suits best when dealing with scrolling text.The most widely used matching algorithms for template matching are the nearest neighbor (NN) algorithm and the RANSAC algorithm. To improve the performance of tracking and to reduce false matching, global matching methods are also employed, such as histogram-based algorithms and MESR-based algorithms.
    - *Tracking with particle filtering*: Particle filtering is a nonlinear method that recursively predicts a system's state founded on its features, available observation models, and sampling. Features such as cumulative histograms, HOG, and projection profiles can be used. The widely used features for this task are edge maps, Harris corners, character strokes, location overlap, and MSERs. Two text blocks are said to belong to the same text if they have identical character strokes and polarity in consecutive frames. A linear classifier combined with interpolation along with features such as temporal distance, overlap ratio, and the edit distance between detected words in current and subsequent frames is used to increase the accuracy in defining the words belong to the same text.

- *Other tracking methods*: Finally, there exist text tracking methods that are similar to object tracking, which are based on continuously adaptive mean shift (CAMSHIFT), optical flow, unscented Kalman filter (UKF), and 3-D rigid motions.
- *Improvisation*: This method uses a feedback loop from text recognition to improve text extraction accuracy and to reduce false positives.

Any method performance depends on the underlying assumptions and the steps involved. Note that depending on the applications, some of these steps can be ignored, or if required, some steps can be included.

*Advantages*

- Effective in separating text segments and background, and hence simple and inexpensive
- Ability to process multilingual text, given multilingual OCR modules [2]

*Disadvantage*

The disadvantage is that it is difficult to integrate several image-processing techniques and to optimize performance measures at every individual step.

### 13.3.3 Training Methods

In contrast to traditional methods, training methods are capable of recognizing characters or words directly using learning and classification approaches. Classification of characters/words plays an important role in training methods. These methods first separate characters from the background and further from each other. The separated words/characters, then match with the words in the dictionary, which is commonly called word spotting [52]. The techniques for word matching employ image features such as HOG apart from advanced classifiers such as random ferns classifiers [53].

*Advantages*

- Less number of steps
- Shows better performance in complex backgrounds
- Effective when dealing with low-resolution videos

*Disadvantage*

Decreased performance while dealing with large text dictionaries and multilingual languages.

## 13.4 Case Studies

### 13.4.1 Static Video Text Detection

In this section a simple and efficient method proposed by Anthimopoulos for caption text extraction from videos is presented in detail. This is a hybrid method that integrates edge-based heuristic algorithms and machine learning algorithms to extract text lines finely by extracting a unique feature set founded on a modified local binary pattern (mLBP). The mLBP feature set defines the distribution of local edge patterns, which actually distinguishes text from the background. Further, the main advantage of this technique is that it can detect text

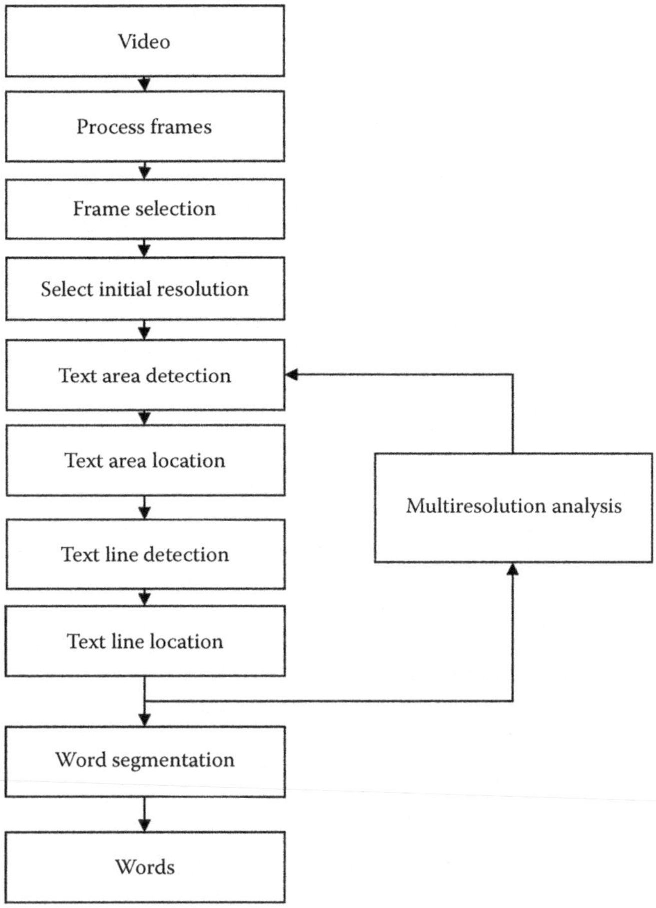


**FIGURE 13.2**
Steps involved in static video text extraction.

regardless of the size and color of characters, as it performs multiresolution analysis. Figure 13.2 demonstrates the block diagram of the Anthimopoulos method of text extraction.

The caption text is assumed to be static, aligned horizontally, and having strong edges. Since processing a single frame is not enough to extract caption text from video, temporal information of multiple frames is employed. The use of multiple frames also helps in removal of noise from the frames by simple averaging (at least 20 frames). One frame out of every five frames of an average video with specific initial resolution is chosen for detection of text, and a Canny edge detector [54] is applied to obtain edge maps. A morphological dilation operator followed by an opening operator is applied to obtain vertical connected components (CC). Coarse text areas are then determined from nonzero CCs.

Fine text lines are separated from text areas by projection analysis or by a hybrid approach that integrates the sliding window method, and SVM classification, which is based on edge feature set. The feature set is obtained from histogram values of mLBP applied to every pixel in the image with a sliding neighborhood of size 3 × 3,

$$\mathrm{mLBP} = \sum c(x,y)2^{n} \tag{13.1}$$

where c(x,y) is a closeness value of pixel centered at coordinates (x,y) and is calculated as follows:

$$c(x,y) = \begin{cases} 0, & D_{ab} < t \\ 1, & D_{ab} \geq t \end{cases} \tag{13.2}$$

where $D_{ab}$ is the absolute difference ($D_{ab}$) between the center pixel and any of its 8 adjacent pixels.

The value of t is chosen to be robust against noise, and small enough to detect sensitive intensity variations in texture. To achieve optimum performance, the value of t is varied adaptively and the relevant mLBP histograms at multiple detail levels are obtained from the following equation:

$$t(i) = -S \cdot \ln \frac{1-i}{M+1} \tag{13.3}$$

where S is the average value of gradient image, i = 1, …, M, and M the number of detail levels.

To ensure low computational complexity and high recall rate, the coarse and fine text line detection algorithms given earlier are performed on multiple low-resolution frames and an analysis algorithm is further used to suppress already detected text lines. As the caption text detection is independent of size and color, this method leads to improved performance for videos with smooth backgrounds. Later, Otsu thresholding method [36] is applied for binarizing the text and background. At this step, it is required to determine if the text is inverse or not using connected component (CC) analysis. The white connected components ($CC_{w}$) and black connected components ($CC_b$), each with vertical lengths greater than eight pixels, are calculated and counted. The text color is then considered as normal (i.e., black) if $CC_b > CC_w$. If not, the text is considered as inverse and is to be inverted to get the normal dark text.

Anthimopoulos also extended this method by employing random forest classifier, which is a parameter-free system and which allows working with small feature sets. This extended version is capable of not only caption text, but also scene text from videos.

### 13.4.2 Scrolling Video Text Detection

Scrolling text is broadly used to display extra data on video frames. To avoid inconvenience and interference, broadcasting rules are defined such that the supplementary text is generally placed and scrolled at the boundaries of the frames (left-to-right/right-to-left/top-to-bottom/bottom-to-top). Given a video, the extracted scrolling text provides important supplemental information for video classification. Further, the information extracted can be recorded for video indexing. In general, algorithms used for the extraction of scrolling caption text employ both temporal and spatial features, as the scrolling text moves over sequential frames. Methods used to detect multioriented text in videos are based on vertical and horizontal histogram distribution. These methods perform poorly in case the video is complex with too many edges or with small text content. Another way is to apply the boundary-growing method along with a Bayesian classifier. This method shows less false positive rate, but with high missing rate. A common drawback of both these approaches is that they cannot determine if the caption text is scrolling or not, as they treat each frame separately. This section presents an efficient

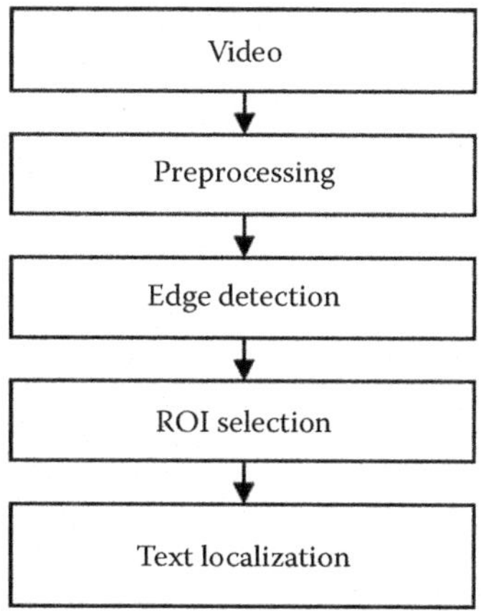


**FIGURE 13.3**
Steps involved in scrolling video text extraction.

algorithm proposed by Hsia et al. that can detect scrolling text from video frames using adaptive temporal differentiation and spatial processing. The algorithm was tested on a huge database of videos, which are captured by catch card on computer from TV in real time and is proven to have no false detections and missed detections. Television is the most prevalent form of entertainment, and the main source of information in the world today. The steps involved are shown in Figure 13.3.

1. *Preprocessing*: This stage is required to reduce computational costs and/or to enhance video frames. Selection of an only intensity component/Y signal (or converting into a gray scale image) makes the system less complex. The enhancement at this stage involves filtering the image for noise removal, and separating text from background information. Though there are a variety of filters such as median/mean filter/Gaussian filter available, Gaussian filters prove effective in removing almost any kind of noise. Median filters are particularly chosen in case of impulse noise.
2. *Edge detection*: This stage is required to enhance text regions. Of many edge detectors available (such as Sobel, Robert, Prewitt, LOG, Canny, Hough), the Roberts edge detector has proved better in differentiating edges in scene and caption texts.
3. *ROI selection*: Since the ROI in the present context is the scrolling text, any non scrolling text region is rejected at this stage. Since it is assumed that the scrolling text appears only at the boundaries, the central region of each frame is rejected first. It is also assumed that a high contrast, and thus high variance, is established to discriminate scrolling text boundary from the background and other boundaries. The temporal differences are expected to be much larger between consecutive frames and are given by

$$D_T = F_t - F_{t-N} \tag{13.4}$$

where
$F_t$ is the current frame
$F_{t-N}$ is the *N*th previous frame

Further, to improve the accuracy of locating the text-related information and to combat the effects of inconsistent scrolling rates of TV channels, an adaptive frame selection speed can be chosen. The number of skipped frames $N$ is inversely proportional to the scrolling text rate and can be estimated by

$$N = \text{FR} \times \text{Scroll Time} \times \text{m} \tag{13.5}$$

where FR is the frame rate, scroll time denotes the time taken by the scrolling text to scroll a specific distance, and m is a constant.

4. *Text localization*: This stage locates and highlights the scrolling text region in a rectangular bounding box. Assuming a black background and white text, and that the text scrolls horizontally, the coordinates of the rectangular bounding box can be obtained easily via scanning the columns one by one and finding the first and last pixels of the text, and marking them as left and right coordinates of the rectangle. Finally the rectangular text box is highlighted with a different color.

## 13.5 Performance Measures

Performance evaluation of video text extraction approaches is also challenging as it is directly related to text detection methods, tracking methods, recognition methods, ground truth data, databases, output formats, and the application. There are several issues related to the complex task of extraction from videos. So far, therefore there is no noticeable comparison of extraction methods and no single text extraction is perfect at detecting text from videos, and only hybrid approaches can be capable of detecting text of different types. Some of the well-known performance metrics are listed here:

1. *Processing time*: It is the average processing time per frame for text detection.
2. *Error rate*: It is the ratio of false detections to total number of true text detections.
3. *Missed detections*: It represents the loss of text detection from the frame.
4. *False alarm/false positives* (*FP*): These are the regions in the frame that are not essentially text characters, but are detected as text. The false alarm rate assesses how many percents of the detected text regions from the video are wrong.
5. *False negatives* (*FN*): These are the regions in the frame that are true text characters, but are not detected.
6. *Precision rate* (*P*): It is the ratio of correctly detected region (CDR) to the sum of correctly detected characters plus false positives.

$$\text{P} = \frac{\text{No. of correctly detected text regions}}{\text{No. of detected text regions}} = \frac{\text{CDR}}{\text{CDR} + \text{FP}} \tag{13.6}$$

The precision rate assesses how many percents of all ground-truth video text regions are properly detected.

7. *Recall rate* (*R*): It is defined as the ratio of the correctly detected text region to the sum of correctly detected characters plus false negatives.

$$R = \frac{\text{No. of correctly detected text regions}}{\text{No. of groundtruth text regions}} = \frac{\text{CDR}}{\text{CDR} + \text{FN}} \quad (13.7)$$

   The recall rate assesses how many percents of the detected videotext regions are correct.

8. *F-score*: F-score is the harmonic mean of precision and recall rates.

$$F = \frac{2 \times P \times R}{P + R} \quad (13.8)$$

9. *Recognition accuracy* (*RA*): It evaluates the percentage of the correctly recognized words.

$$RA = \frac{\text{No. of correctly recoginized words}}{\text{No. of ground truth words}} \quad (13.9)$$

The International Conference on Document Analysis and Recognition (ICDAR), which calls for a Robust Reading Competition every year, uses multiple-object tracking–based metrics such as multiple object tracking accuracy (MOTA), multiple object tracking precision (MOTP), and the average tracking accuracy (ATA) to evaluate the tracking performance of end-to-end video text extraction and recognition. The ICDAR also provides a database of images and videos for a competition. Other commonly used databases are Tan, MSRA-I, MSRA-II, Chars74k, IIIT5K Word, OSTD, Pan, KAIST, NEOCR, etc. These datasets include images and videos captured by a variety of devices to evaluate the performance of text extraction methods developed for multilingual languages.

## 13.6 Challenges and Issues

Text variations with differences in style, size, alignment, and orientation as well as low image contrast and complex background make the automatic text extraction problem tremendously challenging. Uneven illumination: Variations in illuminations cause variations in intensity/color and hence results in degraded images/frames and are the result of capturing devices. Blurred images are the result of defocusing capturing devices, or due to the motion of objects/cameras. Low-quality/low-resolution images are due to compression/decompression applied on videos. Lot of specifications/assumptions are required when dealing with caption text to deal with the lengths, locations, fonts, styles, and aspect ratios of caption text. Languages with large character classes pose difficulty at recognition stage. The detection and recognition rates achieved with the existing methods are often less than 85%, and hence there is much scope for development of advanced image processing, feature extraction, machine learning, pattern recognition, and computer vision algorithms to handle issues related to multioriented text, multilingual text, uneven illuminations, fonts, aspect ratios, noisy backgrounds, and blurred frames. Efficient techniques are required to deal with low-resolution videos. Multiple-instance learning can also be the focus of research. Employing advanced capturing devices, and combining deep learning techniques with multiscale representations, hybrid image-processing operations that combine optimum feature extraction, image enhancement, restoration, segmentation, and tracking techniques can improve the performance of the end-to-end text extraction systems. Further, development of high-performance real-time text extraction and recognition methods is the need of the hour due to increased usage of portable capturing devices.

Challenges in video text extraction arise mainly due to complexity of the background, moving objects, low resolutions, compressed formats, variations in text styles, text distortion, and movement of text across frames. The complexity further increases with real-time videos in terms of accuracy, computations, and speed.

## 13.7 Research Directions

- Multioriented or unaligned scene/embedded text extraction
- Multilingual text extraction systems
- End-to-end text extraction systems
- Real-time text extraction techniques
- Extraction of perspective distorted text
- Development of optimization techniques
- Identification of unique text features
- Investigation of hybrid methods and unique frameworks
- Invention of performance measures
- Fusion of scene and caption texts
- Development of new applications
- Creation of databases for promotion of research

## References

1. Rosenfeld, A., D. Doermann, and D. DeMenthon, eds. *Video Mining*, Vol. 6. Springer Science & Business Media, Springer, NY (2013).
2. Lyu, M. R., J. Song, and M. Cai. A comprehensive method for multilingual video text detection, localization, and extraction. *IEEE Transactions on Circuits and Systems for Video Technology* 15(2) (2005): 243–255.
3. Yin, X.-C., Z.-Y. Zuo, S. Tian, and C.-L. Liu. Text detection, tracking and recognition in video: A comprehensive survey. *IEEE Transactions on Image Processing* 25(6) (2016): 2752–2773.
4. Xiong, Z., X. S. Zhou, Q. Tian, Y. Rui, and T. S. Huang. Semantic retrieval of video. *IEEE Signal Processing Magazine* 23(2) (2006): 18.

5. Hu, W., N. Xie, L. Li, X. Zeng, and S. Maybank. A survey on visual content-based video indexing and retrieval. *IEEE Transactions on Systems, Man, and Cybernetics, Part C (Applications and Reviews)* 41(6) (2011): 797–819.
6. Patel, B. V. and B. B. Meshram. Content based video retrieval systems. arXiv preprint arXiv:1205.1641, *International Journal of UbiComp* (IJU) 3(2) (2012): 13–30.
7. Vijayakumar, V. and R. Nedunchezhian. A study on video data mining. *International Journal of Multimedia Information Retrieval* 1(3) (2012): 153–172.
8. Moxley, E., T. Mei, X.-S. Hua, W.-Y. Ma, and B. S. Manjunath. Automatic video annotation through search and mining. In *2008 IEEE International Conference on Multimedia and Expo*, pp. 685–688. IEEE, Washington, DC (2008).
9. Tseng, V. S., S. Ja-Hwung, J.-H. Huang, and C.-J. Chen. Integrated mining of visual features, speech features, and frequent patterns for semantic video annotation. *IEEE Transactions on Multimedia* 10(2) (2008): 260–267.
10. Dai, K., J. Zhang, and G. Li. Video mining: Concepts, approaches and applications. In *2006 12th International Multi-Media Modelling Conference*, 4pp. IEEE, Washington, DC (2006).
11. Kumar, P. and P. S. Puttaswamy. Moving text line detection and extraction in TV video frames. In *2015 IEEE International Advance Computing Conference (IACC)*, pp. 24–28. IEEE, Washington, DC (2015).
12. Haritaoglu, I. Scene text extraction and translation for handheld devices. In *Proceedings of the 2001 IEEE Computer Society Conference on Computer Vision and Pattern Recognition, (CVPR 2001)*, Vol. 2, p. II-408. IEEE, Washington, DC (2001).
13. Liu, X. A camera phone based currency reader for the visually impaired. In *Proceedings of the 10th International ACM SIGACCESS Conference on Computers and Accessibility*, pp. 305–306. ACM, New York, NY (2008).
14. Shi, X. and X. Yangsheng. A wearable translation robot. In *Proceedings of the 2005 IEEE International Conference on Robotics and Automation*, pp. 4400–4405. IEEE, Washington, DC (2005).
15. Lienhart, R., C. Kuhmunch, and W. Effelsberg. On the detection and recognition of television commercials. In *Proceedings of IEEE International Conference on Multimedia Computing and Systems' 97*, pp. 509–516. IEEE, Washington, DC (1997).
16. Shiratori, H., H. Goto, and H. Kobayashi. An efficient text capture method for moving robots using DCT feature and text tracking. In *18th International Conference on Pattern Recognition (ICPR'06)*, Vol. 2, pp. 1050–1053. IEEE, Washington, DC (2006).
17. Tanaka, M. and H. Goto. Text-tracking wearable camera system for visually-impaired people. In *19th International Conference on Pattern Recognition (ICPR 2008)*, pp. 1–4. IEEE, Washington, DC (2008).
18. Fragoso, V., S. Gauglitz, S. Zamora, J. Kleban, and M. Turk. TranslatAR: A mobile augmented reality translator. In *2011 IEEE Workshop on Applications of Computer Vision (WACV)*, pp. 497–502. IEEE, Washington, DC (2011).
19. Aoki, H., B. Schiele, and A. Pentland. Realtime personal positioning system for a wearable computer. In *Digest of Papers. The Third International Symposium on Wearable Computers*, pp. 37–43. IEEE, Washington, DC (1999).
20. Cui, Y.-T. and Q. Huang. Character extraction of license plates from video. In *Proceedings, 1997 IEEE Computer Society Conference on Computer Vision and Pattern Recognition*, pp. 502–507. IEEE, Washington, DC (1997).
21. Wu, W., X. Chen, and J. Yang. Incremental detection of text on road signs from video with application to a driving assistant system. In *Proceedings of the 12th Annual ACM International Conference on Multimedia*, pp. 852–859. ACM, New York, NY (2004).
22. Coates, A., B. Carpenter, C. Case, S. Satheesh, B. Suresh, T. Wang, D. J. Wu, and A. Y. Ng. Text detection and character recognition in scene images with unsupervised feature learning. In *2011 International Conference on Document Analysis and Recognition*, pp. 440–445. IEEE, Washington, DC (2011).
23. Zhu, Y., J. Sun, and S. Naoi. Recognizing natural scene characters by convolutional neural network and bimodal image enhancement. In *International Workshop on Camera-Based Document Analysis and Recognition*, Beijing, China. pp. 69–82. Springer, Berlin, Heidelberg, 2011.

24. Zhang, H., C. Liu, C. Yang, X. Ding, and K. Q. Wang. An improved scene text extraction method using conditional random field and optical character recognition. In *2011 International Conference on Document Analysis and Recognition*, pp. 708–712. IEEE, Washington, DC (2011).
25. Jung, K. K. I. Kim, and A. K. Jain. Text information extraction in images and video: A survey. *Pattern Recognition* 37(5) (2004): 977–997.
26. Liu, X. and W. Wang. Robustly extracting captions in videos based on stroke-like edges and spatio-temporal analysis. *IEEE Transactions on Multimedia* 14(2) (2012): 482–489.
27. Mancas-Thillou, C. and B. Gosselin. Spatial and color spaces combination for natural scene text extraction. In *2006 International Conference on Image Processing*, pp. 985–988. IEEE, Washington, DC (2006).
28. Kim, K. I., K. Jung, and J. H. Kim. Texture-based approach for text detection in images using support vector machines and continuously adaptive mean shift algorithm. *IEEE Transactions on Pattern Analysis and Machine Intelligence* 25(12) (2003): 1631–1639.
29. Koo, H. I. and D. H. Kim. Scene text detection via connected component clustering and nontext filtering. *IEEE Transactions on Image Processing* 22(6) (2013): 2296–2305.
30. Zhao, X., K.-H. Lin, Y. Fu, Y. Hu, Y. Liu, and T. S. Huang. Text from corners: A novel approach to detect text and caption in videos. *IEEE Transactions on Image Processing* 20(3) (2011): 790–799.
31. Ye, Q. and D. Doermann. Scene text detection via integrated discrimination of component appearance and consensus. In *International Workshop on Camera-Based Document Analysis and Recognition*, pp. 47–59. Springer International Publishing, Switzerland (2013).
32. Garcia, C. and X. Apostolidis. Text detection and segmentation in complex color images. In *Proceedings, 2000 IEEE International Conference on Acoustics, Speech, and Signal Processing (ICASSP'00)*, Vol. 6, pp. 2326–2329. IEEE, Washington, DC (2000).
33. Karatzas, D. and A. Antonacopoulos. Text extraction from Web images based on a split-and-merge segmentation method using color perception. In *Proceedings of the 17th International Conference on Pattern Recognition (ICPR 2004)*, Vol. 2, pp. 634–637. IEEE, Washington, DC (2004).
34. Nikolaou, N. and N. Papamarkos. Color reduction for complex document images. *International Journal of Imaging Systems and Technology* 19(1) (2009): 14–26.
35. Chen, D., J.-M. Odobez, and H. Bourlard. Text detection and recognition in images and video frames. *Pattern Recognition* 37(3) (2004): 595–608.
36. Hanif, S. M., L. Prevost, and P. A. Negri. A cascade detector for text detection in natural scene images. In *19th International Conference on Pattern Recognition (ICPR 2008)*, pp. 1–4. IEEE, Washington, DC (2008).
37. Goto, H. and M. Tanaka. Text-tracking wearable camera system for the blind. In *2009 10th International Conference on Document Analysis and Recognition*, pp. 141–145. IEEE, Washington, DC (2009).
38. Shivakumara, P., T. Q. Phan, and C. L. Tan. New Fourier-statistical features in RGB space for video text detection. *IEEE Transactions on Circuits and Systems for Video Technology* 20(11) (2010): 1520–1532.
39. Li, H., D. Doermann, and O. Kia. Automatic text detection and tracking in digital video. *IEEE Transactions on Image Processing* 9(1) (2000): 147–156.
40. Yin, X.-C., X. Yin, K. Huang, and H.-W. Hao. Robust text detection in natural scene images. *IEEE Transactions on Pattern Analysis and Machine Intelligence* 36(5) (2014): 970–983.
41. Wong, E. K. and M. Chen. A new robust algorithm for video text extraction. *Pattern Recognition* 36(6) (2003): 1397–1406.
42. Jain, A. K. and B. Yu. Automatic text location in images and video frames. In *Proceedings of 14th International Conference on Pattern Recognition*, Vol. 2, pp. 1497–1499. IEEE, Washington, DC (1998).
43. Li, M. and C. Wang. An adaptive text detection approach in images and video frames. In *2008 IEEE International Joint Conference on Neural Networks (IEEE World Congress on Computational Intelligence)*, pp. 72–77. IEEE, Washington, DC (2008).
44. Kim, W. and C. Kim. A new approach for overlay text detection and extraction from complex video scene. *IEEE Transactions on Image Processing* 18(2) (2009): 401–411.

45. Jaderberg, M., A. Vedaldi, and A. Zisserman. Deep features for text spotting. In *European Conference on Computer Vision*, pp. 512–528. Springer International Publishing, Switzerland (2014).
46. Zhiwei, Z., L. Linlin, and T. C. Lim. Edge based binarization for video text images. In *2010 20th International Conference on Pattern Recognition (ICPR)*, pp. 133–136. IEEE (2010).
47. Ferreira, S., V. Garin, and B. Gosselin. A text detection technique applied in the framework of a mobile camera-based application. In *Proceedings of the First International Workshop on Camera-Based Document Analysis and Recognition (CBDAR)*. Seoul, Korea (2005).
48. Lee, S. and J. H. Kim. Integrating multiple character proposals for robust scene text extraction. *Image and Vision Computing* 31(11) (2013): 823–840.
49. Phan, T. Q., P. Shivakumara, B. Su, and C. L. Tan. A gradient vector flow-based method for video character segmentation. In *Eleventh International Conference on Document Analysis and Recognition (ICDAR 2011)*, Beijing, China (2011): 1–5.
50. Novikova, T., O. Barinova, P. Kohli, and V. Lempitsky. Large-lexicon attribute-consistent text recognition in natural images. In *European Conference on Computer Vision*, Florence, Italy. pp. 752–765. Springer-Verlag, Berlin, Heidelberg, 2012.
51. Mishra, A., K. Alahari, and C. V. Jawahar. Scene text recognition using higher order language priors. In *BMVC 2012, 23rd British Machine Vision Conference (BMVA, 2012)*, pp. 1024–1028. IEEE, 2011. University of Surrey, Guildford, Surrey, U.K.
52. Wang, K. and S. Belongie. Word spotting in the wild. In *European Conference on Computer Vision*, Heraklion, Crete, Greece. pp. 591–604. Springer, Berlin, Heidelberg, 2010.
53. Wang, K., B. Babenko, and S. Belongie. End-to-end scene text recognition. In *2011 International Conference on Computer Vision*, pp. 1457–1464. IEEE, 2011.
54. Canny, J. A computational approach to edge detection. *IEEE Transactions on Pattern Analysis and Machine Intelligence* 6 (1986): 679–698.

# 14

## *Deep Learning for Multimedia Content Analysis*

**Nilanjan Dey, Amira S. Ashour, and Gia Nhu Nguyen**

CONTENTS



**ABSTRACT** Multimedia conventional computing techniques depend mainly on the features that are extracted and captured from multimedia content, including text, audio, image, and video data from different domains. Deep learning has a great impact on a variety of applications, including image classification, image clustering, speech recognition, and natural language processing, which effectively apply to multimedia data. Deep learning architectures consist of multiple levels of nonlinear operations. The parameter space searching of deep architectures is a complex task; however, advanced learning algorithms, for instance, the deep belief networks, have lately proposed to tackle the searching process. This chapter discusses the principles and motivations concerning deep learning algorithms such as deep belief networks, restricted Boltzmann machines, and the conventional deep neural network. It deals with the adaptation of deep learning methods to multimedia content analysis ranging from low-level data such as audio and images to high-level semantic data such as natural language. The chapter also addresses challenges and future directions.

**KEY WORDS:** *deep learning, image classification, image clustering, speech recognition, natural language processing, multimedia analysis.*

### 14.1 Introduction

Multimedia content has various forms, including audio, documents, text, statistics, images, graphics/photographs, presentations, video, and software. It has a significant role in assisting computer systems in several applications such as medical, business, educational, and military [1]. In addition, multimedia is valued in offering geographical information and in delivering effective information to individuals with sound effects, animations, films, and motivational materials. Individuals have special needs; for example, blind individuals need audio, while images are necessary for deaf individuals. Audio and video convey information more than that contained in the text. Furthermore, multimedia can convey descriptive information about functional relationships and can offer a consistent experience at any time. Medical publications, videos for surgery, medical images, and endoscopy videos are the several applications for multimedia in the medical domain.

Multimedia content search and analysis can be exhibited as semantic information interactive prediction. Image processing/search, image annotation, and multimedia content analysis applications necessitate the development of several new methods for interaction, processing, analysis, search, and representation with empirical evaluation [2].

Conventional multimedia computing is based on features that are preventive in capturing compound multimedia content including audios, text, and images with domain-specific knowledge. Several techniques were applied to process the multimedia in various applications [3–9]. Current development on deep learning opens a stimulating new era, assigning multimedia computing on a more demanding basis with cross-media interactions and multimodal data modeling using automatically learned representations [10]. Researchers were interested with multimedia content analysis to multimedia content recommendation systems as well as modeling connections between multi-modal data.

Lately, deep learning techniques have been enticing researchers due to their tremendous success in numerous computer vision applications. For multimedia content, particular researchers emphasize on studying multimedia analysis using deep learning techniques for retrieval, detection, segmentation, classification, and tracking [11]. In this chapter, the concept, architectures, and techniques of deep learning methods are addressed. Several applications of deep learning techniques for different types of multimedia are reported. Moreover, future scopes of deep learning in multimedia analysis are presented.

The remaining sections are structured as follows. Section 14.2 presents the concept and techniques of the deep learning approach. Various multimedia applications are included in Section 14.3, where various deep network architectures for effective training and inference approaches in emerging applications of deep learning in multimedia search, management, images segmentation/classification, and retrieval are introduced. The challenges and new perspectives are reported in Section 14.4. Finally, the conclusion is introduced in Section 14.5.

## 14.2 Deep Learning

Deep learning employs various information processing steps in hierarchical constructions for features representation and pattern classification for learning. It is considered a machine learning procedure that is applied in several research areas, including graphical modeling, pattern recognition, neural networks, signal processing, and optimization. Deep learning has a significant role in solving problems that cannot be solved using the artificial intelligence techniques.

The main concept of deep learning is inspired by the artificial neural network techniques [12]. Models with a deep architecture can be originated from many hidden layers of the feed-forward neural networks. The back-propagation algorithm can be applied for weights' learning process of these networks.

Deep learning denotes a broad class of machine learning architectures with the use of many layers of nonlinear information processing. Therefore, based on the way of applying

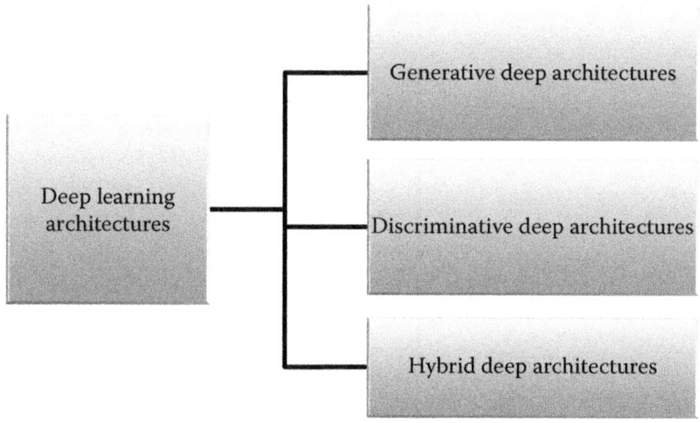


**FIGURE 14.1**
Deep learning architectures.

the architectures and the application under concern, there are three classes of deep learning architectures (Figure 14.1):

1. *Generative deep architectures*: Through this architecture, high-order correlation is captured of the visible data for pattern synthesis/analysis. Moreover, it describes the joint statistical distributions of the observed data and their related classes. In the latter case, the use of Bayes rule can turn this type of architecture into a discriminative one.
2. *Discriminative deep architectures*: The Bayes rule is used to convert the generative deep architectures into discriminative ones. This architecture provides discriminative potential for pattern classification.
3. *Hybrid deep architectures*: This method is applied to discriminate with generative architectures outcomes through better regularization.

Recently, deep learning procedures have achieved state-of-the art performance in several applications. Deep learning has become a significant research topic in both machine learning and computer vision for several multimedia content scenarios. Recently, extensive studies for a variety of deep learning techniques have been developed, such as Boltzmann machines (BM) [13], deep Boltzmann machine (DBM) [14], restricted Boltzmann machines (RBM) [15], deep belief network (DBN) [11], deep neural network (DNN) [16], and the convolutional deep neural network (CNN) [17–20], as illustrated in Figure 14.2.

Some of these techniques are described as follows:

- *Deep convolutional network* [21]: A special category of the feed-forward multilayer neural network. It involves convolutional multiple layers followed by a few fully connected layers.
- *Deep neural network* [22]: A multilayer perceptron with several hidden layers. The weights are entirely linked and are initialized using supervised/unsupervised pretraining method.
- *Boltzmann machine* [23]: A symmetrically linked network, where the stochastic decisions are determined based on the neuron.

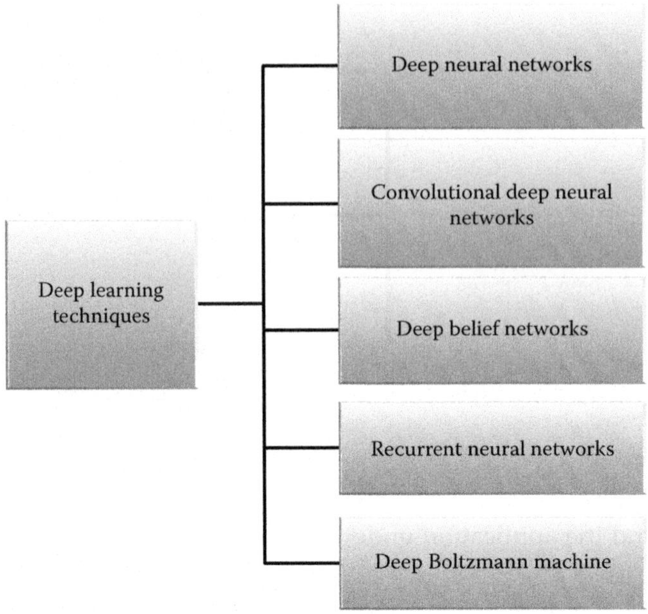


**FIGURE 14.2**
Deep learning techniques.

- *Restricted Boltzmann machine* [24]: A particular BM that consists of a visible unit's layer and a hidden unit's layer without hidden–hidden or visible–visible connections.
- *Deep belief network* [11]: A probabilistic generative scheme consisting of multiple stochastic layers of hidden variables. The first (top) two layers have symmetric/undirected connections. The lower layers receive directed top-down connections from the layer above them.
- *Deep auto-encoder* [25]: represents a DNN whose output is the input data itself.
- *Distributed representation* [26]: A representation produced by the hidden factors' interactions. This type is the foundation of deep learning.

Deep learning techniques can be considered representation learning techniques with several representation levels attained by combining nonlinear simple modules to convert the representation at one level into higher-level representations [27]. These transformations require complex functions to be learned. For example, for an image represented by an array of pixel values:

- A signification of the presence/absence of edges in the trained features at specific locations and orientations in the image occurs in the first layer.
- Detection of the motifs by recognizing specific edges arrangements irrespective of any small variations in the edge positions is performed on the second level.
- Assembling the motifs into larger arrangements corresponding to familiar objects' parts, and succeeding layers would detect objects as arrangements of these parts in the third layer.

The deep learning is characterized by no prior design for the features' layers as they are learned from the data by applying a general-purpose learning technique [28]. Compared to the traditional machine learning techniques, deep learning does not have to resolve the features ahead of time, is fault-tolerant, handles big data easily, and scales well. One of the deep learning potentials is replacing the features with effective algorithms for hierarchical features' extraction and semisupervised/unsupervised feature learning [29]. Since in deep learning, the observed data are produced by the relations of factors corresponding to levels organized in layers, the deep learning algorithms are defined as distributed [30]. For supervised learning, deep learning techniques remove features by interpreting the data into compacted intermediate representations similar to the principal components, and originate layered arrangements for removing the redundancy in representation [31]. Meanwhile, unlabeled data are typically richer than labeled data; hence, deep learning procedures are applied to unsupervised learning applications using deep belief networks.

Deep learning algorithms have various multimedia content applications including automatic speech recognition, image recognition, drug discovery and toxicology, bioinformatics, and customer relationship management.

## 14.3 Multimedia Content Using Deep Learning Applications

Deep learning tolerates the computational representations that consist of multiple processing layers to learn data representations with multiple abstraction levels. These approaches improve dramatically state-of-the-art multimedia content such as visual object recognition, speech recognition, drug discovery, object detection, and genomics. In large datasets, deep learning learns complicated structure using the back-propagation algorithm to designate the change of the internal parameters that are employed to calculate the representation in each layer from the previous layer representation. Generally, CNN has innovations in processing speech, audio, images, and video, whereas recurrent neural networks are promising with sequential data such as speech and text.

Researchers are interested in including deep learning algorithms to handle multimedia content in several domains. In the medical domain, Kalinovsky and Kovalev [32] applied deep learning algorithm and ED-CNN (encoder–decoder CNN) for x-ray chest images segmentation of lungs. The experimental results included a comparison of the proposed approach's segmentation accuracy with the manual segmentation using Dice's score. The comparison established that the average accuracy of the proposed approach was 0.962% with the minimum and maximum Dice's score values of 0.926% and 0.974%, respectively. The results proved that the ED-CNN networks are a promising method for automatic lung segmentation in large-scale applications. Lai [33] investigated three different convolutional network constructions for patch-based segmentation of the hippocampi region in magnetic resonance imaging (MRI) dataset images. The authors applied optimization techniques and deep learning architectures for the diagnosis of Alzheimer's disease.

The CNN deep architecture classification methods are attractive due to their capability to learn mid- and high-level image representations. Thus, in chest radiograph data, Bar et al. [34] examined the deep learning methods' strength for pathology detection. The authors explored the possibility of using a deep learning method based on nonmedical learning, where the CNN was trained. The results proved that the best performance was attained using an arrangement of the extracted features from the CNN and a low-level features set.

Grangier et al. [35] accomplished superior multiple object–class segmentation performance using supervised pretrained convolutional deep neural network. The proposed method was applied for labeling each pixel in an image with one label of a given object–class labels set. Deep learning method based on recursive neural networks was applied in Reference 36 to predict a tree structure for images from multiple modalities. The experimental results proved that the proposed approach achieved very good results for images segmentation and complex image scenes annotation. The recursive neural network algorithm was capable of forecasting hierarchical tree structures.

Wu et al. [37] modeled the deep learning method in a supervised learning framework for image classification. Each image in the tested dataset followed a dual multi-instance assumption. Multiple instance learning with deep learning approaches were exploited with the attempt to learn the association between the object and the annotation.

Sun et al. [38] introduced a hybrid CNN-restricted Boltzmann machine (CNN-RBM) model for learning relational features to compare face similarity. Zhu et al. [39] proposed a new deep learning based face recognition, using face identity-preserving (FIP) features. The FIP features are robust to illumination variations and pose and, in addition, can be used to reconstruct face images. These features were learned by a deep learning model containing feature extraction layers and a reconstruction layer. The results established that the FIP features outperformed the state-of-the-art face recognition techniques. The proposed method improved the classical face recognition schemes by applying them on the proposed reconstructed face images.

For text compressing of handwritten digits/faces, Hinton and Salakhutdinov [40] proved the efficiency of the unsupervised RBM-based deep learning algorithm with pretraining. A feature hierarchy with depth L = 4 was pretrained using the RBMs and afterward fine-tuned for a reconstruction process. This proposed unsupervised learning approach was popular in embedding text documents in a low-dimensional space.

Lee et al. [41] proposed a deep learning scheme to the audio data. The authors showed that since in the case of images learned features correspond to edges, thus for audio data, the features represent phonemes and phones to improve the performance of multiple audio recognition process.

For natural language processing applications, Collobert and Weston [42] applied a deep neural network method for tagging part-of-speech, defining semantic similarity, and semantic roles. For semantic indexing, a deep learning network was described in Reference 43 for learning the documents' binary codes. The first layer of the network signified the word count vector in the document, which considered high-dimensional data. However, the highest layer denoted the trained binary code of the document.

Ranzato et al. [44] trained the deep learning model's parameters based on both unsupervised and supervised data. Thus, it was unnecessary to completely label a large group of data; in addition, the model had prior knowledge to capture relevant labeled information in the data. The results established that deep learning models were superior to shallow learning algorithms to learn compact representations. Analogous to textual data, deep learning can be applied with other data types to extract semantic representations from the input data for further semantic indexing.

For semantic tagging and discriminative tasks, deep learning algorithms can be applied to extract complex nonlinear features from raw data. Afterward, linear approaches can be employed to execute discriminative process using the extracted features as input. The developing of efficient linear models for big data analysis have been broadly investigated in Reference 45. Li et al. [46] explored Microsoft research audio video indexing system (MAVIS) using deep learning based on the artificial neural/networks for speech recognition for searching of audio/video files with speech. For image searching, the authors in [47] employed convolutional neural networks and deep learning for image object recognition using the ImageNet dataset.

## 14.4 Challenges and Future Directions

For multimedia research, it is wholly significant to improve deep networks in order to capture the dependencies between various genres of data, constructing joint deep representation for miscellaneous modalities. Although deep learning has developed enormously over the last years, numerous challenges still remain to be resolved. Some examples of open problems and future directions are as follows:

- The deep learning building blocks are restricted and cannot signify subjective features, since the encoder in the deep autoencoder approach has no hidden layer. For two-layer encoders, unsupervised pretraining is more promising in the future.
- For image sequences to be modeled, the learned invariances set should be extended to comprise transformations and transformations hierarchies.
- Modeling the three-directional structure of scenes to handle occlusions is one of the problems.
- Expanding the proposed methods to be applied with large real-world datasets such as ImageNet is required.
- Spreading deep learning algorithms into applications beyond speech/image recognition will necessitate more software and conceptual innovations.
- Future development in vision from systems that are trained and integrate ConvNets with RNNs using reinforcement learning will be considered. Systems combining reinforcement learning and deep learning will be extended to outperform passive vision systems for classification applications, to produce inspiring results with various video scenarios.
- Deep learning–based natural language applications has large potential over the next fewyears. It is expected that systems using the RNNs are developed to comprehend sentences and/or whole documents.
- Multimedia content includes big data, thus deep learning challenges in big data analytics should be considered. Typically, deep learning requests additional investigation, precisely, streaming data learning, high-dimensionaldata handling, models scalability, and distributed computing. For nonstationary data, incremental learning is a challenging aspect with fast-moving input data and streaming. Such data analysis is valuable in monitoring applications, such as fraud detection. It is significant to adjust deep learning for managing data streaming with large amounts of continuous input data.

- Some deep learning procedures deal with high-dimensional data and accordingly become extremely computationally exclusive. This leads to slow learning procedure associated with the deep-layered hierarchy for learning data abstractions and demonstrations from a lower- to higher-level layer.
- Optimization algorithms can be used to support the deep learning techniques for various multimedia applications.

Deep learning will achieve more advancement in the near future due to its very few requirements; thus it can handle simply large amount of existing computation and data. Consequently, the current learning architectures and algorithms for deep neural networks will accelerate this advancement.

Eventually, the foremost progress in artificial intelligence will occur through systems that gather complex reasoning with representation learning, though simple reasoning and deep learning were used for handwriting-/speech-recognition. Innovative paradigms are desirable to exchange symbolic expressions by operations on large vectors.

## 14.5 Conclusions

In the medical imaging domain, machine learning is applied for image segmentation, registration, and fusion, as well as for computer-aided diagnosis, image-guided therapy, image database retrieval, and image annotation. Recently, deep learning approaches represent machine learning set of algorithms that try to automatically learn multiple representations and abstraction levels to obtain sense from the data. This in turn requires examining and understanding deep learning approaches' features to be able to refine and to apply the deep learning approaches in an appropriate way.

Deep learning has a potential advantage of supplying a solution for data analysis and learning problems found in massive volumes of input data in contrast to feature engineering systems and conventional machine learning. More precisely, it supports automatic extraction for complex massive data representations of unsupervised data. This makes it a valued tool for big data analysis that involves analysis from huge raw data collections that is commonly unsupervised and uncategorized. Deep learning can be employed effectively for analyzing massive data volumes, data tagging, semantic indexing, information retrieval, and discriminative tasks like classification and prediction.

Feature hierarchies lead to time efficiency and space decomposition of inputs that can be convenient in numerous tasks, including denoising, classification, and compression. Deep learning solution aims to learning feature hierarchies in order to solve a structure of simple shallow problems. Then, in each step, deep learning approaches learn a novel features level to acquire new visions into the input data distribution.

## References

1. Costello, V. *Multimedia Foundations: Core Concepts for Digital Design*. CRC Press, Boca Raton, FL, 2016.
2. Kennedy, L. Advanced techniques for multimedia search: Leveraging cues from content and structure. Doctoral dissertation, Columbia University, New York, 2009.

3. Roy, P., Goswami, S., Chakraborty, S., Azar, A.T., and Dey, N. Image segmentation using rough set theory: A review. *International Journal of Rough Sets and Data Analysis (IJRSDA), IGI Global*, 1(2):62–74, 2017.
4. Pal, G., Acharjee, S., Rudrapaul, D., Ashour, A.S., and Dey, N. Video segmentation using minimum ratio similarity measurement. *International Journal of Image Mining (Inderscience)*, 1(1):87–110, 2015.
5. Samanta, S., Dey, N., Das, P., Acharjee, S., and Chaudhuri, S.S. Multilevel threshold based gray scale image segmentation using cuckoo search. In *International Conference on Emerging Trends in Electrical, Communication and Information Technologies (ICECIT)*, December 12–23, 2012.
6. Bose, S., Mukherjee, A., Madhulika, S.C., Samanta, S., and Dey, N. Parallel image segmentation using multi-threading and K-means algorithm. In *2013 IEEE International Conference on Computational Intelligence and Computing Research (ICCIC)*, Madurai, India, December 26–28, 2013.
7. Dey, N. and Ashour, A. (eds.) *Classification and Clustering in Biomedical Signal Processing*. Advances in Bioinformatics and Biomedical Engineering (ABBE). IGI Book Series, 2016.
8. Karaa, W.B.A., Ashour, A.S., Sassi, D.B., Roy, P., Kausar, N., and Dey, N. MEDLINE text mining: An enhancement genetic algorithm based approach for document clustering. *Applications of Intelligent Optimization in Biology and Medicine: Current Trends and Open Problems*. 2015.
9. Chakraborty, S., Dey, N., Samanta, S., Ashour, A.S., and Balas, V.E. Firefly algorithm for optimized non-rigid demons registration, *Bio-Inspired Computation and Applications in Image Processing*, Yang, X.S. and Papa, J.P. eds., 2016.
10. Mohamed, A., Dahl, G., and Hinton, G. Acoustic modeling using deep belief networks. *IEEE Transactions Audio, Speech, & Language Processing*, 20(1):14–22, January 2012.
11. Hinton, G.E., Osindero, S., and Teh, Y.W. A fast learning algorithm for deep belief nets. *Neural Computation*, 18(7):1527–1554, 2006.
12. Deng, L. A tutorial survey of architectures, algorithms, and applications for deep learning. *APSIPA Transactions on Signal and Information Processing*, 3:e2, 2014.
13. Ackley, D.H., Hinton, G.E., and Sejnowski, T.J. A learning algorithm for Boltzmann machines. *Cognitive Science*, 9(1):147–169, 1985.
14. Salakhutdinov, R. and Hinton, G.E. Deep Boltzmann machines. In *Artificial Intelligence and Statistics Conference*, pp. 448–455, 2009.
15. Salakhutdinov, R., Mnih, A., and Hinton, G.E. Restricted Boltzmann machines for collaborative filtering. In *International Conference on Machine Learning*, pp. 791–798, 2007.
16. Hinton, G., Deng, L., Yu, D., Dahl, G.E., Mohamed, A.-R., Jaitly, N., Senior, A. et al. Deep neural networks for acoustic modeling in speech recognition: The shared views of four research groups. *Signal Processing Magazine, IEEE*, 29(6):82–97, 2012.
17. Ciresan, D.C., Giusti, A., Gambardella, L.M., and Schmidhuber, J. Deep neural networks segment neuronal membranes in electron microscopy images. In *NIPS*, pp. 2852–2860, 2012.
18. Dean, J., Corrado, G., Monga, R., Chen, K., Devin, M., Le, Q.V., Mao, M. Z. et al. Large scale distributed deep networks. In *NIPS*, pp. 1232–1240, 2012.
19. Krizhevsky, A., Sutskever, I., and Hinton, G.E. Imagenet classification with deep convolutional neural networks. In *NIPS*, pp. 1106–1114, 2012.
20. LeCun, Y., Bottou, L., Bengio, Y., and Haffner, P. Gradient-based learning applied to document recognition. *Proceedings of the IEEE*, 86(11):2278–2324, 1998.
21. Razavian, A.S., Azizpour, H., Sullivan, J., and Carlsson, S. CNN features off-the-shelf: An astounding baseline for recognition. In *2014 IEEE Conference on Computer Vision and Pattern Recognition Workshops (CVPRW)*. IEEE, pp. 512–519, 2014.
22. Denil, M., Bazzani, L., Larochelle, H., and de Freitas, N. Learning where to attend with deep architectures for image tracking. *Neural Computation*, 24(8):2151–2184, 2012.
23. Larochelle, H. and Hinton, G.E. Learning to combine foveal glimpses with a third-order Boltzmann machine. In *Advances in Neural Information Processing Systems*, Vol. 23, pp. 1243–1251, 2010.

24. Dahl, G., Ranzato, M., Mohamed, A., and Hinton, G. Phone recognition with the mean-covariance restricted Boltzmann machine. In *Proceedings of NIPS*, Vol. 23, pp. 469–477, 2010.
25. Cho, K., van Merrienboer, B., Gulcehre, C., Bougares, F., Schwenk, H., and Bengio, Y. Learning phrase representations using RNN encoderdecoder for statistical machine translation. In *Proceedings of the Empiricial Methods in Natural Language Processing (EMNLP 2014)*, October 2014.
26. Mnih, A. and Hinton G. A scalable hierarchical distributed language model. In *Proceedings of NIPS*, pp. 1081–1088, 2008.
27. Knowles-Barley, S., Jones, T.R., Morgan, J., Lee, D., Kasthuri, N., Lichtman, J.W., and Pfister, H. Deep learning for the connectome. In *GPU Technology Conference*, 2014.
28. Larochelle, H., Bengio, Y., Louradour, J., and Lamblin, P. Exploring strategies for training deep neural networks. *Journal of Machine Learning Research*, 10:1–40, 2009.
29. Song, H.A. and Lee, S.Y. Hierarchical representation using NMF. In *Neural Information Processing*. Lectures Notes in Computer Sciences 8226. Springer, Berlin, Germany, pp. 466–473, 2013.
30. Bengio, Y., Courville, A., and Vincent, P. Representation learning: A review and new perspectives. *IEEE Transactions on Pattern Analysis and Machine Intelligence*, 35 (8):1798–1828, 2013.
31. Deng, L. and Yu, D. Deep learning: Methods and applications (PDF). *Foundations and Trends in Signal Processing*. 7(3–4):1–199, 2014.
32. Kalinovsky, A. and Kovalev, V. Lung image segmentation using deep learning methods and convolutional neural networks. In *XIII International Conference on Pattern Recognition and Information Processing*, October 2016.
33. Liao, S., Gao, Y., Oto, A., and Shen, D. Representation learning: A unified deep learning framework for automatic prostate MR segmentation. In *International Conference on Medical Image Computing and Computer-Assisted Intervention*, Springer, Berlin/Heidelberg, pp. 254–261, 2013.
34. Bar, Y., Diamant, I., Wolf, L., and Greenspan, H. Deep learning with non-medical training used for chest pathology identification. In *SPIE Medical Imaging*. International Society for Optics and Photonics, pp. 94140V, March 20, 2015.
35. Grangier, D., Bottou, L., and Collobert, R. *Deep convolutional networks for scene parsing*. In *ICML Deep Learning Workshop*, Montreal, Quebec, Canada, 2009.
36. Socher, R., Lin, C.C., Ng, A., and Manning, C. Parsing natural scenes and natural language with recursive neural networks. In *Proceedings of the 28th International Conference on Machine Learning*, Omnipress, pp. 129–136, 2011.
37. Wu, J., Yu, Y., Huang, C., and Yu, K. Deep multiple instance learning for image classification and auto-annotation. In *2015 IEEE Conference on Computer Vision and Pattern Recognition (CVPR)*, IEEE, pp. 3460–3469, June 7, 2015.
38. Sun, Y., Wang, X., and Tang, X. *Hybrid deep learning for face verification*. In *ICCV*, 2013.
39. Zhu, Z., Luo, P., Wang, X., and Tang, X. Deep learning identity-preserving face space. In *Proceedings of the IEEE International Conference on Computer Vision*, pp. 113–120, 2013.
40. Hinton, G. and Salakhutdinov, R. Reducing the dimensionality of data with neural networks. *Science*, 313(5786):504–507, 2006.
41. Lee, H., Pham, P., Largman, Y., and Ng, A. Unsupervised feature learning for audio classification using convolutional deep belief networks. In *Advances in Neural Information Processing Systems (NIPS)*, Vancouver, British Columbia, Canada, pp. 1096–1104, 2009.
42. Collobert, R. and Weston, J. A unified architecture for natural language processing: Deep neural networks with multitask learning. In *Proceedings of International Conference on Machine Learning (ICML)*, Helsinki, Finland, pp. 160–167, 2008.
43. Hinton, G. and Salakhutdinov, R. Discovering binary codes for documents by learning deep generative models. *Topics in Cognitive Science*, 3(1):74–91, 2011.

44. Ranzato, M. and Szummer, M. Semi-supervised learning of compact document representations with deep networks. In *Proceedings of the 25th International Conference on Machine Learning*, ACM, pp. 792–799, 2008.
45. National Research Council. *Frontiers in Massive Data Analysis*. The National Academies Press, Washington, DC, 2013.
46. Li, G., Zhu, H., Cheng, G., Thambiratnam, K., Chitsaz, B., Yu, D., and Seide, F. Context-dependent deep neural networks for audio indexing of real-life data. In *2012 IEEE Spoken Language Technology Workshop (SLT)*, IEEE, pp. 143–148, 2012.
47. Krizhevsky, A., Sutskever, I., and Hinton, G. Imagenet classification with deep convolutional neural networks. In *Advances in Neural Information Processing Systems*, Vol. 25. Curran Associates, Inc., pp. 1106–1114, 2012.

# 15

# *Video-Image-Text Content Mining*

**Adjan Abosolaiman**

**CONTENTS**



**ABSTRACT** Nowadays, videos and images contain text data that indicate to useful information for indexing, retrieval, automatic annotation, and structuring of images. The extraction of this information can be executed by several phases on a digital video. This chapter explains in detail different phases for text extraction and approaches used in every phase. The phases are preprocessing and segmentation, detection, localization, tracking, extraction, and recognition, respectively. In addition, the chapter discusses several suitable techniques according to the video type and phase. Mechanically, when these techniques have been applied, the text in video sequences will be extracted to provide useful information about their contents. Furthermore, this chapter aims at extraction of text information from video (such as news videos) and multimodal mining from the same.

**KEY WORDS:** *video text detection, extraction, mining, localization and segmentation.*

## 15.1 Introduction

Since the last decade, with the growth of digital devices, videos have become the language of communication and learning in our daily life. Nowadays, most people transmit and share their status or new ideas through designing videos on social media applications as Snapchat or Facebook and others. Videos have become the platform of communications with communities and distance learning. As a result, there is a continuous increment in the number of video databases on the Internet and there is no doubt that there is an increasing need for algorithms to classify, analyze, detect, and recognize information that is displayed in videos, in order to reach the targeted video or retrieve it.

Video text extraction and mining is still a field in its infancy, compared with other types of data [1]. This chapter aims to present varied approaches for detecting text and the steps of extracting and mining information from videos.

## 15.2 Related Work

Many text detection and extraction approaches have been proposed in the past several years. We found that video text extraction or mining can be conducted according to several aspects. The first aspect is that videos differ in what is contained. Vijayakumar and Nedunchezhian discuss two types of videos for mining that are related to content structure [1]. They divided videos that are used in our daily life into videos with some content structure as in news/movies videos, and videos without any content structure as in sports videos [1–3].

In this decade, if we consider videos, we will discover that videos can be divided into several types. Videos' editors express their ideas in different ways. Some editors design their videos with graphic text and scene text. Graphic text or text overlay is the video text that editors add mechanically in videos such as news and sport videos. As Zhang and Chang [4] discuss, these are video texts or scene texts that are recorded by cameras. Scene texts are video texts embedded in the real-world objects or scenes, for example, videos that include car license number, street name, number/name on the back of a football player, and so on. The third type of videos, based on explanation of incidents as in news scene or lectures, are called speech- and text-based videos, which Kate and Waghmare experimented with [5] to extract knowledge. Figure 15.1 illustrates the differentiation among the video types.

However, all these three mentioned types that are found in researches do not preclude the presence of mixtures among them. So we have to choose the right approach that is suitable to the video in order to extract information. The next section will discuss all the phases of different approaches for video text retrieval system, based on the video contents and the researchers' vision as shown in Figure 15.2.

Generally, text extraction in videos can be summarized in the following stages.

(a) (b) (c) (d) (e)

**FIGURE 15.1**
Examples of video text types: (a) graphic text, (b) scene text, (c) news scene, (d) news tape scene, (e) lecture scene.

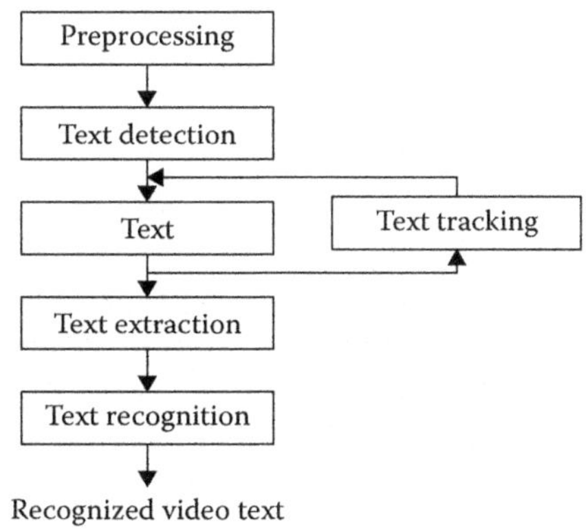


**FIGURE 15.2**
Video text retrieval system.

### 15.2.1 Preprocessing Stage

The first stage is preprocessing videos and transforming them from nonrelational data into a relational dataset by partitioning the video contents into a set of hierarchical manageable units such as clusters, subclusters, subregions, shots or objects, frames or video object planes, and regions that contain text [6]. The most common use of partitioning is shot-based and object-based [3]. The shot-based retrieval approach is used with videos that consist of content structure. The object-based retrieval approach is used with videos that do not consist of any content structure. Inexpensively, vector features can be extracted from compressed MPEG videos.

As (www.eetimes.com) states that MPEG compression file format comes from "moving picture experts group [7]." MPEG comprises fully encoding only key frames that applied with JPEG algorithm (Joint Photographic Experts Group) to estimate the changes of motion between these key frames. MPEG generally, breaks up videos into group of pictures by selecting a snap shot every four frames. Consequently, it minimizes the number of pictures in a video, that causes a significant reduction in bits required to describe the video. As a result of that compression ratios above 100:1 are common. The scheme of MPEG encoder is very complex and extremely sensitive in estimating place of motion.

### 15.2.2 Detection Stage

The second stage is text detection (finding regions in a video frame that contain text). Zhang and Chang [4] distinguish the video text detection and recognition as key components for the video analysis and retrieval system. Due to complex backgrounds and low resolution of videos and various colors, sizes, styles, and alignments of text, text detection and extraction are still challenging stages [8]. There are three methods to perform text detecting: (1) Method of edge-based extraction, which can quickly locate the text area. If video frame contains strong edge information, it will have relatively high accuracy. (2) Method that is texture-based. This method is more versatile than the others, but it usually performs Fast Fourier Transforms (FFTs) that compute the temporal object boundary pixels to detect peaks in Fourier domain, and also wavelet transform, which is time consuming [9]. (3) Method based on time-domain characteristics, which uses the appearance and disappearance of video caption text to detect text area. This method is occasionally used when camera shakes strongly and text moves frequently [8].

### 15.2.3 Localization Stage

The next stage is text localization. It is the process of grouping text location in the image into text instances and then generating a set of tight bounding boxes around all text instances [8,10]. This step is considered as a segmentation process that segments localized text into lines that merge into single line text, after which the line is segmented according to characters using connected component analysis based method.

### 15.2.4 Tracking Stage

The stage of text tracking is responsible for reducing the processing time of text localization and maintaining the integrity of position across adjacent frames. Although the accurate location of text in an image can be determined by bounding boxes, the text still needs to be segmented from the background to simplify its recognition [10].

### 15.2.5 Extraction Stage

The text extraction step can be performed by two groups of methods. The first group is based on color differentiation methods, and the other on stroke-based methods. The former group assumes that the text pixels are sharply apparent. This means there are colors with consideration to contrast between text and background pixels. Besides the validity of the assumption, there is another difficulty of this approach, that is, the color polarity of text (light or dark) that must be determined. On the other hand, the stroke-based methods run some filters to output only those pixels that are likely on the strokes to the final results, such as the asymmetric filter, the four-direction character extraction filter [6], and the topographical feature mask [8]. These filters are created to enhance the stripe (i.e., stroke-like) shapes and to ignore others. However, the intersection of strokes may also be ignored due to the lack of stripe shape. Shivakumara et al. [11] suggested the wavelet transform technique and gradient features for text extraction, respectively [8].

Segmentation of a video text into text lines and characters is a critical stage toward recognition. The segmentation procedure is a challenging task due to some problems that arise in video text. There are many difficulties encountered in the segmentation; these include the overlapping of words and adjacent text, the variance in the skew angle between lines, characters, or even along the same text line, lines, and touching characters.

### 15.2.6 Recognition Stage

The last stage is text recognition, which performs optical character recognition (OCR) on the binarized text image, after the segmentation stage. Sometimes, binarization step on color/gray level images is rejected in favor of applying OCR.

Since any video consists of a series of images either fixed or mobile with a sound, optical character recognition can be used to separate textual data from each frame, and extract audio using automatic speech recognition (ASR). We will address both techniques in the next section.

#### *15.2.6.1 OCR Technology*

OCR (optical character recognition) is a technique that converts different types of scanned images that are captured by digital camera of documents, (PDF files, sales receipts, mail, handwritten, typewritten, or any number of printed records) into searchable and editable data. It is widely used for extracting textual metadata, that is, machine-encoded text. (www.abbyy.com) states that the recognized document by OCR looks like the original. Therefore, these textual data can be used in machine processes such as machine translation, text to speech, and text mining. The OCR software allows saving a lot of time and effort spent in creating and processing and repurchasing various documents. OCR is a field of research in pattern recognition, artificial intelligence, and computer vision [12,13].

OCR technology has been increasingly applied to paper-intensive industry. It deals with more complex image environment in the real world, for instance, low-resolution, heavy-noise, complicated backgrounds, paper skew, degraded images, picture distortion, disturbed by grid & lines, text image consisting of special fonts, glossary words, symbols, and so on [13].

#### *15.2.6.2 Templates in OCR*

Basically, OCR technique uses templates that need to be designed for template matching on scanned image. Originally, these templates are images of letters. An OCR could use font-free templates, but if there is a need to be more specific then a font-based OCR can be designed. Then the image should be first converted into a binary image that contains either black or white. Hence, all the colors' values in the image are converted to 0 and 1. This transformation to binary image basically removes noise from the image, as also the third component of the image is removed, which provides ease of reading [13].

After binarizing, the image is negated, that is, if the image contains the white letter on black background, then it converts to the black letter on white background, and vice versa. Basically, this makes the process of matching much easier between the letter template and the letter in the input image. This is how the font-based template is generated for OCR (Figure 15.3).

The corr2 is a function in MATLAB that is used for the matching of two images A and B, when A and B are the matrices and vectors of the same size. Corr2 (A,B) computes the correlation coefficient between A and B [13].

#### *15.2.6.3 ASR Technology*

Automatic speech recognition (ASR) can be defined as an independent, computer-driven transcription that transforms spoken language to readable text in real time [14]. ASR is the technology that allows a computer to recognize and identify the words that a person speaks in a video and convert it to written text [15].

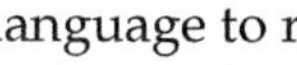


**FIGURE 15.3**
Image of binarizing as illustrated by Gaikwad, H. et al. (From Gaikwad, H. et al., *Int. J. Eng. Res. Technol.*, 2(3), 2013.)

#### *15.2.6.4 Video Mining*

Video mining is used to reveal and characterize interesting patterns among video data [1]. Applying video data mining algorithms aims to find correlations and patterns previously unknown from large video database [16]. Many video mining approaches have been proposed for extracting valuable knowledge from video database. In general, it can be classified into three categories: spatial pattern detection, video association mining, and video clustering and classification [3]. Video association rule mining is one of the serious problems in video data mining. Commonly, mining association rule is a straightforward extension of association rule mining in transaction databases.

## 15.3 Video Text Extraction and Mining Approaches

### 15.3.1 Video Text Extraction Based on Structured Contents

In the first phase in video preprocessing, the video is transformed from the nonstructured data into a structured form. Next, it is converted into a temporal sequence database. A video can be observed as a sequence of images bounded with spatial and temporal properties. These are simply segmented into shots. These shots can be chosen by identifying the boundary between shots automatically. Usually, a shot is represented by a key frame, which is then used for extracting features such as color, audio, text, motion, and objects. Vijayakumar and Nedunchezhian [1] studied object feature, which involves in the video semantic concept and event. So, they conducted their work on object-based featured extraction. They manually generated the video sequence.

Videos which have structured contents are those that are recited according to their displaying structure, customary among all organizations as in events news/movies videos. Vijayakumar and Nedunchezhian [1] studied an algorithm of mining that uses Best-N frequent pattern in a video sequence. The searchers preprocessed the videos to access the semantic information from the video. Firstly, they converted the video data from the nonstructured data into a structured form. Then they transform video frames into a temporal video sequence database. Finally, this temporal video sequence database is undergone to extract the frequent subsequences by applying mining algorithm. They proposed a new algorithm VidApriori and modified it to generate frequent patterns. Frequent pattern generation plays a real role in mining of association rules. The modified VidApriori algorithm can extensively decrease the search space.

### 15.3.2 Video Text Extraction Based on Images

This experiment of Zhang and Chang [4] described a system to detect and extract the textual information in digital video, as illustrated in Figure 15.4. The research proposed different approaches from the previous. The system used a multiple hypothesis testing approach,

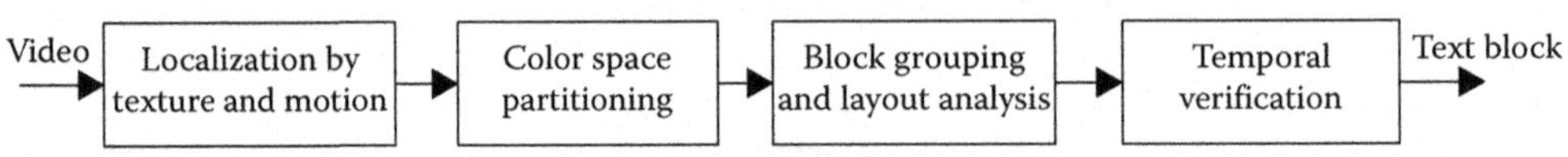


**FIGURE 15.4**
Zhang and Chang system flowchart. (From Zhang, D. and Chang, S.F., Accurate overlay text extraction for digital video analysis.)

which is the region-of-interests (ROI). It probably includes the overlay texts that are decomposed into several hypothetical binary images, applying color space partitioning. Then, a grouping algorithm is conducted to group the recognized character blocks into text lines in each binary image. If the grouped text lines' layout conforms to the verification rules, detected text regions produce the bounding boxes of these grouped blocks. Finally, motion verification is used to increase the accuracy of text extraction. ROI localization is achieved using compression of MPEG video format, in order to achieve real-time speed. The suggested method led to achieving impressive results with average recall 96.9% and precision 71.6% in testing on digital News videos.

### 15.3.3 Video Text Extraction Based on Scene

Scene text that can be recorded by digital cameras can be recognized. One of the most basic and significantly important steps in text recognition is scene text character (STC) prediction. It can be judged as a multiclass classification among a set of text character categories. Yi [17] designed a feature to represent the STC structure, by combining multiple feature descriptors, coding/pooling schemes, and learning models.

He suggested a scene text extraction framework, which is applied to four examples of scenarios as seen in Figure 15.5: (1) reading printed labels in grocery package to recognize hand-held objects; (2) combining naturally captured scene image by camera with car detection to localize license plate; (3) reading indicative signage to aid navigation in indoor environments; and (4) combining with object tracking to extract scene text in video-based scene that looks natural [17].

To overcome the mentioned challenges, they needed to concentrate on two problems. Firstly, how to typically design text layout and structure so that it can be distinguished from nontext background outliers, that is, scene text region detection. Secondly, how to

**FIGURE 15.5**
Examples of: (a) hand-held objects; (b) car license plate; (c) indoor environments signage; (d) scene text in video-based.

design the structure of scene text characters so that the class of a given character can be predicted precisely, that is, scene text character prediction.

To solve the two discussed problems that Yi studied, his framework of scene text extraction is divided into two functional modules in the baseline solutions [18,19]: scene text detection and scene text recognition. Both modules scene text detection and scene text recognition produce text detector and text recognizer, respectively, and they have introduced into two research topics. Scene text detection is to localize the image regions that is containing text and strings and nominate most background interferences. Some detection methods are used in text or strings segmentation in detected text regions into independent characters for recognition. Scene text recognition method is used to convert image-based text strings in detected text regions into readable ASCII codes. Figure 15.6 clarifies the flowchart of the proposed framework.

Yi evaluated the performance of his proposed framework on three benchmark datasets of scene images. First is the ICDAR-2003 [20], followed by the ICDAR-2011 [21], which are collected for robust reading competitions, and annotated text regions. ICDAR 2003 database contains about 500 scene images and 2258 text regions in total. In their experiments, some scene images do not contain text and some contain only a single character. Thus, 487 scene images are used to evaluate performance. The range of image size is from 640 × 480 to 1600 × 1200.

The ICDAR-2011 Robust Reading dataset contains 484 scene images with 848 ground truth text regions in total; 229 images are used for training and 255 images for testing in ICDAR 2011 Robust Reading competition. They evaluated the framework on all the images that contain two character members or more. The range of image size is from 422 × 102 to 3888 × 2592. The proposed framework is applied on the mentioned datasets to localize text. The localization processes are applied on each scene image and its inverse image, and the results are combined for calculating the localized text regions.

The second database is Born-digital images and broadcast video images [22], also used to evaluate Yi's framework. He used electrical documents with colorful captions and illustrations in Born-digital images database, which mostly exist in web pages, book covers, and posters. Born-digital image has higher occurrences frequency of text and smaller character sizes than scene image. A dataset of Born-digital images is released for ICDAR-2011 robust reading competition [22]. It contains 420 Born-digital images with ground truth text regions. The average image size is about 352 × 200.

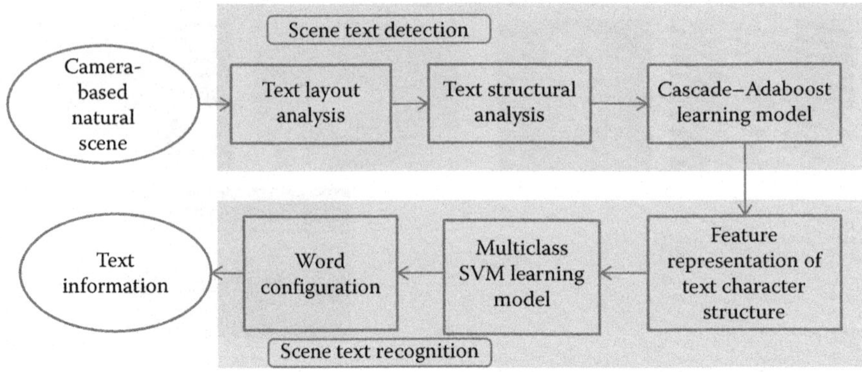


**FIGURE 15.6**
The flowchart of Yi proposed scene text extraction framework.

The third database is Street View Text dataset [23], collected from Google street view. This dataset is captured from outdoor environments with illumination variations. In this database, most text characters have low resolutions and are embedded into complex background outliers. It consists of 350 scene images and 900 ground truth text regions in total. In Street View dataset, backgrounds are interfering and more complex, which are generated due to more false positive detections. So in this dataset, the precision is much lower than that in ICDAR Robust Reading dataset.

Yi evaluated his proposed approach of text localization in images/videos by conducting a comparison between a group of detected text regions and ground truth text regions that are manually labeled. He defined the overlaps between detected regions and ground truth regions as hit regions, which mean the detections are correct. Then he defined the area of a text region as the number of pixels in the region.

Yi defined the precision as the ratio between the area of hit regions and the area of the detected regions; this is used to sense the amount of false positives in the detected regions. He also defined recall as the ratio between the area of hit regions and the area of the ground truth regions; this is used to sense the amount of missing detections in the ground truth regions. Then they are combined by harmonic mean to obtain f-measure as in Equation 15.1:

$$\text{f-measure} = \frac{2^{*}\,\text{Precision}^{*}\,\text{Recall}}{\text{Precision} + \text{Recall}} \tag{15.1}$$

#### *15.3.3.1 Evaluating Scene Text Detection by Different Layout Analysis on ICDAR Datasets*

At first, Yi evaluated the performance of text layout analysis without looking at the structural feature analysis. His experimental results on the Robust Reading dataset indicated that the combination of color-based partition and adjacent character grouping (CA) reaches the highest precision and recall, as illustrated in Table 15.1. Color uniformity plays as a stronger indicator in most of the cases to differentiate the connected components of text characters from the background that surrounds characters [17].

**TABLE 15.1**

The Comparison between Yi's Framework and the Text Localization Algorithms Presented in [15,24,25] on the Robust Reading Dataset

| Method | Precision | Recall | f-Measure |
|---|---|---|---|
| **C. Yi** | **0.73** | **0.67** | **0.66** |
| B. Epshtein | 0.73 | 0.60 | 0.66 |
| H. Becker | 0.62 | 0.67 | 0.62 |
| **C. Yi** | **0.71** | **0.62** | **0.62** |
| A. Chen | 0.60 | 0.60 | 0.58 |
| Ashida | 0.55 | 0.46 | 0.50 |
| H. W. David | 0.44 | 0.46 | 0.45 |
| Wolf | 0.30 | 0.44 | 0.35 |
| Q. Zhu | 0.33 | 0.40 | 0.33 |
| Wolf | 0.30 | 0.44 | 0.35 |
| J. Kim | 0.22 | 0.28 | 0.22 |
| Todoran | 0.19 | 0.18 | 0.18 |
| N. Ezaki | 0.18 | 0.36 | 0.22 |

**FIGURE 15.7**
Some example results of text localization in the ICDAR-2011 dataset, where the text regions are marked by cyan boxes with Yi's algorithm.

**TABLE 15.2**
The Results of ICDAR-2011 Robust Reading Competition on Scene Text Localization (%)

| Method | Precision | Recall | f-Measure |
|---|---|---|---|
| **C. Yi (2012)** | **81.00** | **72.00** | **71.00** |
| Kim | 62.47 | 82.9871 | 71.28 |
| **C. Yi (2011)** | **58.09** | **67.22** | **62.32** |
| TH-TextLoc | 57.68 | 66.97 | 61.98 |
| Neumann | 52.54 | 68.93 | 59.63 |
| TDM_IACS | 53.52 | 63.52 | 58.09 |
| LIP6-Retin | 50.07 | 62.97 | 55.78 |
| KAISTAIPR | 44.57 | 59.67 | 51.03 |

*Source:* Yi, C., Text extraction from natural scene: Methodology and application, 2014; Shahab, A. et al., ICDAR 2011 robust reading competition, in *International Conference on Document Analysis and Recognition*, pp. 1491–1496, 2011. Yi's proposed framework won second place.

Moreover, Yi [17] achieves the highest precision and recall when he combines layout analysis and structural analysis using ICDAR-2003 dataset for performance evaluation. The precision goes up to 0.73 and recall to 0.67. Therefore, f-measure increased to 0.66. In addition, in 2012, Yi won the second place in the evaluation of the proposed algorithm applied on ICDAR-2011 (Figure 15.7) Robust Reading competition comparing with other scene text localization (%). It rose to 81.00, 72.00, and 71.00 for precision, recall, and f-measure, respectively, as shown in Table 15.2.

#### *15.3.3.2 Evaluating Scene Text Detection on Born-Digital Images and Video Images Dataset*

Further, Yi evaluated his framework for extracting text information from Born-digital images and broadcast video images dataset. Born-digital images are electrical documents with colorful captions that mostly exist in web pages, for example, book covers and posters. Besides, Born-digital images have higher frequency of occurrences of text and smaller character sizes than scene images (Figure 15.8). In most video images, text can be as titles

**FIGURE 15.8**
Example results of text localization in born-digital images presented by Yi.

and captions to indicate the content of television program. The text is distributed on the top or bottom of the image. The strings and characters also have the features of bigram color uniformity, character alignment, and stroke width consistency. Most text information in broadcast video images differs from scene images. Generally, they have fewer background interferences and pattern variations. The following figure characterizes some examples of text localization in Born-digital images and Figure 15.9 in broadcast video images [17].

Although Yi's framework is able to reveal most text information in natural scene that has complex background, there are still some challenging situations that his framework cannot handle correctly. The main reason of this failure is that these situations do not accept his assumption of text layout and structure. The following images depict some examples which his method cannot handle the localization of the text information because of some trammels. Firstly, text may be is very small size, less than three character members, non-uniform colors or fade, overexposure, or occlusions caused by other objects such as wire mesh [17] (Figure 15.10).

**FIGURE 15.9**
Example results of text localization in broadcast video images presented by Yi.

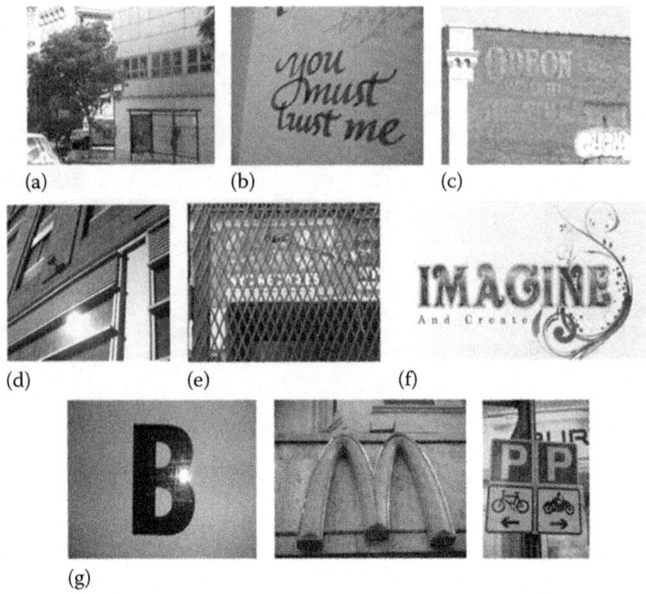


**FIGURE 15.10**
Some examples of images where Yi's method fails: (a) too small size, (b) ligature, (c) blur and fade, (d) overexposure, (e) wire mesh, (f) colored characters, (g) text string containing less than three characters.

### 15.3.4 Video Text Extraction Based on Image Shots and Speech

As discussed, videos consist of images and speech that complete each other, such as videos for lectures and news. First, in the preprocessing stage, Kate and Waghmare [5] transform the video into key frames and extract the audio and text using OCR and ASR, addressed previously. The following step outputs a summary introducing key points of the video, by making use of text and audio extracted from the video. Then, this summary is used for videos' grouping and indexing. This in turn will improve the user's efficiency to quickly review the targeted material. This will make users go through only information that they need. Nevertheless, in the videos, text may vary in dimension, style, orientation, background, contrast, and variations in rhythm, volume of and noise in speech, and the differentiating between the key speeches and dispensable other sounds that are recorded as well. All this makes data extraction extremely challenging.

## 15.4 Conclusion

In this chapter, we presented different approaches for different video types to retrieve textual data as well as audio data, automatically. In fact, videos over the World Wide Web can contain the textual data. However, they may have different color, style, size, and may have a plain or natural background or be handwritten. Similar to textual data, audio keywords also may have different volume, tempo, and all sorts of noise mixed with it. Regarding the previously mentioned researches, for retrieving data, either textual or audio, we can automatically index and group large video archives. These approaches can save video producers time in writing and describing video-related information manually. This will be beneficial for users, as they will not need to go through long and boring videos to reach the video that they looking for, and will get only the videos they need. However, this chapter still does not distinguish a new method for extracting and mining text. It is just presenting different methods that use for different videos types.

## References


1. Vijayakumar, V. and Nedunchezhian, R., 2011. Mining best-N frequent patterns in a video sequence. *International Journal on Computer Science and Engineering*, 3(11), 3525.
2. Ma, Y.F., Lu, L., Zhang, H.J., and Li, M., December 2002. A user attention model for video summarization. In *Proceedings of the 10th ACM International Conference on Multimedia*, Columbia University, New York, NY: ACM, pp. 533–542.
3. Zhu, X., Wu, X., Elmagarmid, A.K., Feng, Z., and Wu, L., 2005. Video data mining: Semantic indexing and event detection from the association perspective. *IEEE Transactions on Knowledge and Data Engineering*, 17(5), 665–677.
4. Zhang, D., Tseng, B.L., and Chang, S.F., 2003, August. Accurate overlay text extraction for digital video analysis. In *Information Technology: Research and Education, 2003. Proceedings. ITRE2003. International Conference on* (pp. 233–237). IEEE.
5. Kate, L.S. and Waghmare, M.M., 2014. A Survey on Content based Video Retrieval Using Speech and Text information. *International Journal of Science and Research* (IJSR), 3(11), 1152–1154.
6. Sato, T., Kanade, T., Hughes, E.K., and Smith, M.A., January 1998. Video OCR for digital news archive. In *Proceedings of the 1998 IEEE International Workshop on Content-Based Access of Image and Video Database*, Carnegie Mellon University, pp. 52–60. IEEE.
7. Andrew, D., 2016. An overview of video compression algorithms, [Online], Available: http://www.eetimes.com/document.asp?doc_id=1275884 [August 17, 2016].
8. Vinod, H.C., Niranjan, S.K., and Anoop, G.L., 2013. Detection, extraction and segmentation of video text in complex background. *International Journal on Advanced Computer Theory and Engineering*, 5, 117–123.
9. Töreyin, B.U., Dedeoğlu, Y., and Cetin, A.E., September 2005. Wavelet based real-time smoke detection in video. In *Proceedings of the 13th European Signal Processing Conference*, Bilkent University, Ankara, Turkey, 06800, pp. 1–4. IEEE.
10. Jung, K., Kim, K.I., and Jain, A.K., 2004. Text information extraction in images and video: A survey. *Pattern Recognition*, 37(5), 977–997.
11. Shivakumara, P., Phan, T.Q., and Tan, C.L., 2009. A robust wavelet transform based technique for video text detection. In *Proceedings of the 2009 10th International Conference on Document Analysis and Recognition*. National University of Singapore, IEEE.
12. Deshmukh Bhagyashri, D., November 2014. Review on content based video lecture retrieval. IJRET: *International Journal of Research in Engineering and Technology*, 3(11), Pune University, India.
13. Gaikwad, H., Hapase, A., Kelkar, C., and Khairnar, N., March 2013. News video segmentation and categorization using text extraction technique. *International Journal of Engineering Research and Technology*, 2(3), 2278-0181. ESRSA Publications.
14. Stuckless, R., 1994. Developments in real-time speech-to-text communication for people with impaired hearing. *Communication Access for People with Hearing Loss*, Ross, M. ed. Baltimore, MD: York Press, pp. 197–226.
15. Pore, A.R. and Sahu, A., 2014. Survey on speech recognization techniques. (IJCSIT) *International Journal of Computer Science and Information Technologies*, 5(2), 2263–2267. Amravati, Maharashtra, India.

16. Oh, J., Lee, J., and Hwang, S., 2005. Video data mining: Current status and challenges. In *Encyclopedia of Data Warehousing and Mining*, Wang, J. ed. Idea Group Inc. and IRM Press, University of Bridgeport, Bridgeport, CT.
17. Yi, C., 2014. Text extraction from natural scene: Methodology and application.
18. Zhang, J. and Kasturi, R., 2008. Extraction of text objects in video documents: Recent progress. In *IAPR International Workshop on Document Analysis Systems*, Nara, Japan, pp. 5–17.
19. Jung, K., Kim, K., and Jain, A., 2004. Text information extraction in images and videos: A survey. *Pattern Recognition*, 5, 977–997.
20. ICDAR., 2003. http://algoval.essex.ac.uk/icdar/Datasets.html.
21. ICDAR., 2011. http://robustreading.opendfki.de/wiki/SceneText.
22. ICDAR., 2011. http://www.cvc.uab.es/icdar2011competition/.
23. Wang, K., 2010. http://vision.ucsd.edu/~kai/svt/.
24. Lucas, S., 2005. ICDAR 2005 text locating competition results. In *Proceedings of the International Conference on Document Analysis and Recognition*, pp. 80–84.
25. Lucas, S. et al., 2003. ICDAR 2003 robust reading competition. In *Proceedings of the International Conference on Document Analysis and Recognition*.
26. Shahab, A., Shafait, F., and Dengel, A., 2011. ICDAR 2011 robust reading competition. In *Proceedings of the International Conference on Document Analysis and Recognition*, pp. 1491–1496.

# Index

**D**

## N



## O



## P

## Q



## R



## S



## T

**W**



**X**